My Vietnam

Chicago to Southeast Asia

Vernon J Gregory

ISBN: 978-1-7357174-0-1

Description

It is a compelling, adrenaline charged, powerful description in the words of a United States combat veterans experience in the war in Vietnam. Growing up in the small town of Riverdale, Illinois on the south side of Chicago opportunities abound. It was a great carefree life until July 7, 1970 when Vernon Gregory received his draft notice. Here Vernon is planning his future, working for a local excavator making top dollar, playing Semi-Pro baseball and planning on getting married. This life story will hopefully reveal the trials and tribulations of the emotional and moral consequences of combat in war. In this book Vernon Gregory talks about many of his missions and the emotions that have followed him through life. The transition from combat to civilian life involves more than taking off the uniform. Vernon wants veterans of all wars to know they are not alone and help is available for them and their loved ones through the Veterans Administration and Veteran Centers. This book may help significant others understand what their loved one has been subjected to. Vernon would like to honor our fallen brothers and sisters in arms and to those whose wounds are not visible. They were normal kids put into an abnormal situation and expected to return normal.

Preface

My Vietnam is about my life prior to my military service, some of my combat missions and its affect on my life.

Prior to my military experience, the United States was going through tumultuous times. President Kennedy was assassinated in 1963, there were riots for social justice and protests against the Vietnam War. None of this affected me as I was growing up. I played sports and worked one month a year on a farm in the summer.

Then on July 7, 1970 my induction notice came and shook my life forever. I left much behind and took combat very seriously, but not without fear. My combat experience become etched in my brain and took its toll on me and my family.

Writing this book was to help me deal with my experiences and to hopefully help loved ones of combat veterans and those who have not been in the military understand the trials and tribulations of combat and its affects.

CONTENT

CHAPTER 1

Chicago 1970

I was 19 years old when the nation was under the regime of our thirty-seventh President Richard Nixon. This particular moment in time was two years after the riotous Democratic Convention in Chicago. Our country was in a state of turmoil. Bad decisions became a way of life for our political leaders. People were angry with the actions of the government and challenged their policies. Rallies were held for different causes and a lot of people died protesting their beliefs. The people of the 1970s were not afraid to participate in dangerous controversies. They were a society obsessed with making changes. The issues of government interested every scale of social classes. Musicians wrote memorable songs about current events and the movie stars began to take active roles in politics. There were a lot of important events that documented history. This is the year Charles Manson was indicted. The movie M.A.S.H was released. Paul McCartney announced he was leaving the Beatles. The nation celebrated the first annual event of Earth Day. Apollo 13 failed at the attempt of the third moon mission. A bill signed into law lowered the voting age from 21 to 18.

I grew up in the muscle car generation, almost experiencing the Woodstock Concert. I survived the sex, drug and bell bottom revolution of the 60s. My time was invested working for a local excavating company, playing semi-pro baseball and steadily dating a girl for three years. I lived with my Mother and Grandparents in Riverdale, Illinois.

The house I grew up in

The town of Riverdale had a population of 15,806. Riverdale was a small community dominated by white Americans. The house in which I lived was located on the southwest corner of 139th Street and Emerald Avenue. Emerald Avenue is a one block dead end street located in the northwest corner of the town. There were sixteen houses on my block. The majority of the homes were owned by strict Italian families. My grandparents were of Italian descent and spoke the language fluently. The neighbors living within my vicinity were concerned about maintaining a high standard of living. They never allowed any individual to disrupt the harmony of the neighborhood. There was one particular rule on the block which was enforced. The rule pertained to children loitering the street after dark. The neighborhood curfew went into affect the moment the street lights came on. Children were supposed to be in the company of their parents when night time arrived. Any seen violating their curfew were told to go home. The neighbors would not tolerate children ignoring their requests. The neighbors thought being uncooperative was a form of disrespect. The actions of a disrespectful child were reported to the parents. Phone calls about inappropriate conduct was considered a family embarrassment. The children realized the consequences that resulted from a unfavorable telephone conversation. The punishment was disciplinary action.

I was raised in a prejudice environment. I was led to believe that different races should not intermingle with one another. My association with different races was almost nonexistent throughout my childhood. African Americans, Latinos and Hispanics were equally discriminated against because they shared a troublesome reputation. Problems always seemed to follow these groups wherever they would congregate. The negative projection of these ethnicities was due to the results of a few troubled individuals, whether it was psychological or by poverty. The stereotyping of these races was influenced by the way our parents, relatives and friends expressed themselves. Television and the motion picture industry added to our ignorance by portraying certain races to be inferior and categorized them negatively.

I attended a Catholic Grade School that had a white student body of approximately six hundred and forty students. The only contact I had with other races was observing them commuting between different towns. The laws of our government prohibited anyone from denying members of a different ethnicity the right to work, receive an education and be provided with public transportation. The laws of desegregation were beginning to make a change in our society by the time I turned fifteen. Minority races wanted more opportunities in the working world. They wanted to move into safe neighborhoods and attend better schools. The goals of the minority race was to have the opportunity to prove themselves as equals.

I encountered a variety of different races while attending Mendel Catholic College Preparatory High School. My four years of high school taught me to interact with people of different cultures and religious backgrounds. My Catholic education emphasized the ugliness associated with racism. I learned that each individual was a child of God and judged by merit not their skin color. My teachers stressed the importance of equality and the need to respect people as human beings. The students at Mendel had respect for one another but there was always a feeling of racial tension. Classmates were always conscious of what they said to the other person because they did not want to offend anyone. The education I received had many conflicting beliefs between myself, the Catholic Church, my family and the ideals of what the white suburban ethnic groups practiced. The neighborhood was a stable environment compared to the acts of violence certain families faced 3 miles away from my house.

Everything seemed to be going smooth for me until the day of July 7, 1970. My world began to crumble the moment the mailman delivered the mail at 11:30am Tuesday morning. I slowly began to weaken as I read a draft notice addressed to me by the United States Government. The notice stated that I was to report for duty on Wednesday, July 29, 1970. I was number one hundred eighty-two in the draft lottery. This was supposed to be the year in which the draft was not to go over ninety. I was scared, confused and definitely not ready to leave home. I became a very angry young man. I wanted to know why someone would want to do this to me. I thought I had no enemies. I never caused any trouble or had a police record. I was a drug free individual who wanted to be left alone so I could progress with my life. I had three cousins who served and fought in Vietnam. They were lucky to have survived the combat in the jungle. The only reminder my cousins had of the war were wounds from hostile fire and a Purple Heart. I noticed a big change in my cousins when they returned home. I did not know what gave them a different outlook on life but I realized they acted differently when they returned home.

I was very uncertain about participating in the war after I talked to my three cousins. I realized the odds were against me and there was a good chance I might not survive the war. My cousin Butch understood my concerns and assured me that I would be all right. He told me to take one day at a time and focus on my training. Butch advised me to never volunteer for unfamiliar duties unless I understood the specific procedures of a particular situation. Butch served with pride in Vietnam as a Marine in 1967. He was wounded in the war with shrapnel from a mortar round that struck him, tearing out part of his lung, stomach and spleen.

I tried to think of several schemes that would alter my chances of serving in the army. My first thought was to flee the country and hide out in Canada. Canada was the place to go to dodge the war and make a new life. Extradition did not exist in Canada. The officials of the United States Government could not come in the country and interfere in matters across the Canadian border. A lot of sons of prestigious political families became draft dodgers. The topic of draft dodgers became a serious issue in the 1970s. I can remember a particular song written by a member of Credence Clearwater Revival Band named Dan Fogerty. The title of his song was called "The Fortunate One". This anti-war song criticized wealthy families for finding ways to prevent their sons from serving in the war.

My friends encouraged me to fail the Army's physical examination by devising unique tactics to become rejected. The service was very selective in recruiting individuals during the time of war. The military would not allow a person to enlist in the service if they could not meet certain physical requirements. Any type of physical impairment would make it impossible for an individual to successfully complete their six weeks of basic training. The Army would be legally responsible if a physically unfit person became injured. The ideas of my friends had great potential. Their concepts might have prevented me from becoming selected for military duty. I politely declined all of the offers I received out of fear of the physical pain that I would have to endure. I felt confident that I could come up with a plan that did not involve so much pain such as the ones my friends had presented.

I thought the Army would not allow the only male child to serve in a war. Certain male children were given deferments because their families needed someone to carry on the family name. The process of obtaining a deferment from the Armed Forces was complicated. The law required a parent to submit three letters to three different governmental offices. The offices that considered the decision were the State Senator, Governor and the local Representative. The changes of receiving a determent was almost impossible because it took over a year to process the paperwork. The reality of my misfortune was beginning to change my way of thinking. I thought the service would allow me to remain with my mother so I could take care of her. I tried to explain these feelings to my family but they would not accept my reasons. My mother and grandfather told me it was my duty to serve my country as a United States citizen. I could not understand why my family wanted me to participate in an unpopular war that did not have the support of the general population. I realized being drafted was the law but it did not mean I had to like the decision. My grandfather told me if nothing else, the Army would teach me how to walk straight.

My life reached the lowest turning point imaginable on Thursday, July 9, 1970. I was forced to give up a lifetime dream of playing professional baseball.

1969

I was further devastated as I read my letter of acceptance from the Chicago White Sox. I felt God was handing me the career opportunity of a lifetime. I realized a chance like this happens only once in a persons life. I wanted the fair chance of a private tryout in August at Comiskey Park. I knew I would have made the team because I possessed a natural talent for baseball. I hated the idea of having to sacrifice a career with the Chicago White Sox. My draft notice forced me to decline the baseball offer.

A party was held in my honor three days before my induction into the United States Army. A childhood friend named Bruce arranged this wild event. Bruce lived seven houses away from me in the middle of the block. I considered Bruce to be a lucky individual because the draft did not reach him. He was spared.

Bruce took me to a southside Chicago bar located in a town called Roseland. The bar was established in an area nicknamed Bum Town. This particular bar had the reputation of serving under age kids alcohol. I was not a drinker. Bruce did all of the ordering because I was not familiar with alcohol. I drank slow gin fizzes from eight o'clock in the evening to one thirty in the morning. I did not realize the amount of alcohol I drank until I became sick. Bruce had to pull his motorcycle over to the side of the road so I could vomit. We drove to a local forest preserve because we both realized we were too drunk to be out on the road. We found a wooden picnic bench and decided to go to sleep there. We were awaken by a Cook County Sheriff Police Officer at six o'clock in the morning. The officer told us we were trespassing on forest preserve property and endangering our lives by sleeping in a preserved area. The officer warned us never to come back and allowed us to be on our way. The party ended when Bruce extended his hand and wished me the best in front of my house. I was too drunk to walk into my house. I did not have the energy to find my room and get into bed. The alcohol forced me to pass out in my back yard. I spent the entire day sleeping in a lawn chair because I was too hung over to move to a comfortable spot. My Aunt Teresa happened to stop by in the afternoon. The first thing my Aunt did was walk over to me. I was lying there with my sunglasses on. She said, "I have what will cure you." She walked into the house and came out with a glass of tomato juice. I told her I did not want it but she said, "Drink this, you will feel better." I drank it and I'll be damned I felt better.

I spent the last day home with my girlfriend Candy. This was the saddest and the shortest day of my life. I avoided looking at the clock because I knew what tomorrow would bring. Tomorrow morning would be the day I began a new life. 5 a.m. would be the hour I had to leave everything I loved and cared about behind. The war was tearing me away from my family, girlfriend, dog, job and all of my worldly possessions. Things were happening too fast. My life was becoming unstable. There were no guarantees of anything anymore. All my hard work and planning for a better future slipped right through my hands. I wondered if I would be able to adjust in my new world.

CHAPTER 2

My D-Day

In my mind, July 29, 1970 was my D-Day. This particular D-Day did not refer to the World War II invasion of Normandy. This day meant devastation day to me. I said goodbye to my mother, grandparents, girlfriend and Candys family. This was a very emotional experience. Everyone tried their best to hold back their tears. My girlfriend cried the hardest. I hated the thought of leaving my girl behind. I wanted to marry her not leave her. Candy was a beautiful girl. This would be the first time I would be separated from her. It seemed strange to part because we experienced everything together. She was the only girl I knew who enjoyed hunting and fishing as much as I did. Candy promised she would write on a daily basis. She vowed to wait for me until I returned from the service. Candy reassured me that I would be her boyfriend. I wondered if Candy would keep her promises. You try hard to remember everything to say because you do not want anything to be left unsaid.

Nobody knew at this point how drastic my life was about to change. I had to report to the person in charge of our group at the Harvey, Illinois train station at 5:30am. Our group leader took role call and handed us our train passes. We took the Illinois Central train into the City of Chicago. The train dropped our group off on Van Buren Street. We walked the rest of the way to the Army Induction Center at Wells Street and New Orleans Street. Once we reached the induction center we were given a physical. The exams were very embarrassing due to the presence of women walking back and forth with paperwork. We were told as a group to strip down to our underwear. During one particular exam there were about fifty men standing in a single file line. We were instructed to drop our underwear down to our ankles while a doctor checked our penis to see if we had been circumcised. Then we were told to turn around, spread our feet apart, bend over and spread our butt cheeks with our hands. The doctor walked past each recruit inspecting everyone's posterior opening to the digestive tract. To this day I do not know what the doctor was looking for.

After our physical we were given an aptitude test, a mechanical ability written test and a common sense test. There were a total of two hundred and ninety men taking these tests. The tests were rather diversified and there were only five men who achieved high scores. I was one of the five men who scored exceptionally well. The five of us were escorted from the main group and put into a room where there was a Sergeant. The Sergeant separated us individually and brought each one of us before a Captain. The Captain was sitting behind a desk in an enclosed room within the same building.

The Captain told me I scored extremely high on my test and that I had the potential the armed services were looking for. He gave me some options to seriously consider and wanted an answer before our conversation ended. The Captain told me I qualified for West Point Cadet Academy in New York. If I attended this school I would learn the required military strategies and troop coordination on how to defeat the enemy in a war situation. Regular term would require five years of active duty and three years of inactive duty. The second position I qualified for was the ASA (Army Security Agency). The duties of this agency was to support the field armies and national defense effort with electronic communications security. The required stay in service for the ASA was four years of active duty and two years of inactive duty. The third position I qualified for was OCS (Officer Candidate School). This school would prepare an individual for decision making and managing groups of men of lower rank. The serving time for Officer Candidate School required four years of active duty and two years of inactive duty.

I politely turned down all three offers up to this point. I told the Captain I felt the military lifestyle was not for me and I did not want to remain in the service for more than my required stay as a draftee for two years. The Captain responded by saying he had something of a shorter term that might be to my liking. He asked me if I would like to attend Warrant Officer School. A Warrant Officer is an officer without authority. A Warrant Officer made more money than a Non-commissioned Officer and had the benefits of a commissioned officer. Warrant Officers were trained to fly helicopters in combat and supportive roles. The amount of time required for Warrant Officer training was three years active duty and three years of inactive duty. I turned the Captain down politely for the fourth time. The Captain paused and asked me if I would be interested in attending NCO (Non Commissioned Officer School). A non-commissioned officer is a subordinate officer appointed from the enlisted men as a Sergeant. They delegate orders to smaller groups of men and manage all work activities. The required stay for going to NCO School was three years of active duty and three years of inactive duty. This is the fifth time I turned the Captain down.

The Sergeant walked into the room and asked the Captain if he made any progress. The look of disgust was on the Captains face as he looked at the Sergeant and shook his head in disbelief. The Captain and Sergeant both told me I was missing the opportunity of a lifetime. They believed the education and the training I would receive would be beneficial to me in the work environment after I was discharged from the service. The Sergeant told me he wished he could have had some of the choices I was offered. The Sergeant commented that he had to start at the bottom of the ladder and work his way up to his present rank. It took this particular Sergeant five years to attain his rank. The Sergeant was aggravated with my over all attitude because he knew I could have become a high ranking official within a short period of time if I would have accepted one of the Captains offers. The Captains voice changed from a pleasant to a very hostile one because I kept refusing his five offers. The Captain told me that if I changed my mind before leaving for Fort Lewis I could report back to him and he would take care of the necessary paperwork.

I did not refuse the five career offers because I wanted to be a smartass, nor did I refuse the Captain because I felt I was too good for the service. This was not my attitude. In fact, I was almost afraid to turn the Captain down. The real reason I did was because I just wanted to go home. My girlfriend Candy was another reason I did not want to stay past two years in the service. I had dreams of making a life with her. I wanted to marry her and raise a family.

I spoke to the four guys who scored extremely high on their tests and none of them received the same five offers that had been offered to me. I discovered that three men turned the Captain down the same way I did. There was only one who accepted the Captains offer. This particular man chose to attend OCS school. Our buses arrived and it was time for us to leave for the airport. It seemed like it took forever to reach O'Hare International Airport. I do not know what was worse, the time being stuck in traffic or the anxieties that arise before boarding an airplane. I had never flown on a airplane before and I did not know what type of surprises to expect. I thought about all kinds of ways to ditch my group. I thought of refusing to go. I thought about hiding in a room where I would be left behind. I thought about Canada. I realized the penalties there were to pay for dodging the Army. The specific clause written in any draft notice states that willful failure to report at the place and hour of the day named in the order subjects the violator to a fine and imprisonment. Prison did not scare me. The only thing that kept me from doing anything irrational was remembering my family values. I did not want to bring embarrassment or any type of dishonor to my family.

I found flight 193 to be very exciting. The only thing I did not like about flying was that you reached your destination too fast. It took us exactly four and a half hours to reach the airport in Seattle, Washington. After we disembarked the plane we realized that we were going to be shuttled to Fort Lewis, Washington. Our bus ride to Fort Lewis lasted about an hour. Even though I did not like what was happening to me, I somehow enjoyed the scenery along the way. It was such a change from Chicago where there was nothing but tall skyscrapers. The temperature was in the high 70s with no smog in the air. Everywhere you looked you could see the landscapes of mountains. Evergreen trees were prolific in the state of Washington and the scent of pine permeated the air. I really enjoyed the scenery of another part of the country because I was an outdoors person who had an appreciation for nature.

When I got off the bus, I saw a lot of men dressed in dark olive uniforms and wearing Smokey the Bear hats. One man in particular named Sergeant Brown came up to us and started giving us orders. This is where our military training began. Sergeant Brown told us stand straight with the tips of our shoes on the front edge of the yellow line and to space ourselves from one another. Then we were to extend our right arm and touch the shoulder of the person next to us for proper spacing. This was the way we would line up every time we would have formation. The Sergeant commenced to say that he is now our father, mother, brother and sister. If anyone had a problem about anything we were instructed to report the matter directly to him. He informed us that we need to pay close attention to the instructions we receive, do exactly what our instructor says and learn as fast as possible because we had a one hundred percent chance of going to Vietnam. The Sergeants job was to teach us survival in a hostile environment. He told us that if we got killed in Vietnam that he would hear about it and not feel any responsible obligation because he gave us the necessary knowledge and tools to survive. After the Sergeant finished speaking, he ordered us to leave our personal belongings where we stood and leave the area by marching. We marched to a building where we were sworn into the United States Army. I could see from this point the Army was not going to play games with us nor were they going to tolerate any bullshit.

My first night I slept in a barrack with over a hundred people I was not familiar with. This was my first experience sleeping in the same room with different ethnicities. Many thoughts rushed through my head. I wondered If I was going to be killed in my sleep or awakened by being beaten. I felt confident that no one had weapons. We had to throw what the Army called contraband in a basket when we were sworn in. I felt that if I had to fight, I would be able to defend myself because I was a street fighter and boxer from the south side of Chicago. That night I only got about an hour and a half of sleep. I did not socialize or try to make friends because my fears were very real. I minded my own business and followed directions as they were given to me.

The next day we were measured and fitted for our uniforms. Our personal belongings were then boxed and sent home. At that point we had no association with anything else but military items. In the same week we learned many lessons such as personal hygiene. Each group shaved and brushed their teeth, then moved on to the next area where we waited in line to get our hair cut. Every room was partitioned off so you could not see what was going on. You entered the building at one end and exited on the opposite side. In the barber shop, the barbers waited until everyone was seated, spun you around to at look pictures on the wall and asked how you would like your hair cut. It did not matter which style you chose because everyone received the house special. The house special was a military crew cut. On the preceding days we received dental check ups and numerous amounts of medical injections. The rules of the mess hall were that soldiers were given ten minutes to eat in total silence.

We are now dressed and look like soldiers. My basic training is beginning with Company C 3rd Battalion 2nd Brigade. Company C had 4 platoons and each platoon had its own barrack. I was assigned to a barrack and a bunk. In this platoon barrack we were divided into four squads.

I realized that my days of privacy were over. Every move I made from now on was done as a group. It was very humiliating for me when it came time to use the washroom. I did not like the fact of having to shit before a live audience. There were no walls that separated the twenty commodes in the mens room, nor was there any privacy when it came to using the urinals because they were large troughs lining the walls. The showers were set up in a large room with no partitions, similar to the ones used in high schools. I was a very modest person and had a terrible time dealing with all of this. I even had a problem with our sleeping arrangements because I was not used to sleeping in a room filled with guys. Our sleeping quarters were not partitioned either. There were bunks spread out evenly three feet to my right and left sides. Each barrack had a person that was responsible for fire watch. This person walked the floor all night and remained alert in case of trouble. These factors caused a tremendous amount of tension in my life because I was not accustomed to this.

The first month of military life was the most difficult. I hated being away from home and I missed my girlfriend. I would have liked nothing better then to grab my fishing pole and summon the dog I grew up with to go fishing with me. Any type of activity would have been better than hanging around and having visions of myself being killed. Every day that went by was a day closer for me to leave for Vietnam. These uneasy feelings caused me to suffer a nervous condition… anxiety. I had to be rushed to the base emergency room because I was experiencing shortness of breath and my fists were clenched so tight that a crow bar could not have pried them open. The physician on duty administered a shot of valium to relax me and prescribed two pills for me to take later. The doctor assured me that I would be alright. Sergeant Brown approached me about my medical incident and wanted to know if I needed to talk. At that time, I felt he could do nothing for me and reneged on his offer.

We had the opportunity to request counseling from the chaplain, so I decided to take them up on their offer. Drill Sergeant Brown said he would set up a meeting and let me know when it would be held. Three days later the Sergeant informed the company that meetings with the Chaplain were going to be arranged due to so many requests. At this time I knew I was not alone. I realized there were other people as scared as I was or least had similar concerns. I went to the chapel the night our Sergeant told us the Chaplain was ready to see our group. I never received a chance to speak to the chaplain because I was told the meetings were falling behind schedule. Instead I had to talk to Sergeant Brown first. He wanted to know what I needed to talk to the Chaplain about. I addressed the Sergeant in a sincere manner and cried as I talked. I told him how I wanted to get out of the Army and explained to him why he should discharge me. I let him know I was the only male child in the family left to carry on the family name. I informed him that I needed to return home so I could take care of my mother and grandparents. I expressed my concerns about having a family member die when I was not around.

The sergeant tried to help me in his own way, I suppose. He explained to me that death is a natural part of life and we should not let our fears force us into irrational actions or thinking. Sergeant Brown told me the possibility of getting out of the service on a technicality could be arranged, but the initial paper work would never be completed before I left for Vietnam. The Sergeant advised me to strive for self control and learn everything I could before leaving for Vietnam. Sergeant Brown denied me the right to see the Chaplain and told me if I needed to talk further that he would make himself available.

I realized there was no possible way I would get out of this. I remembered what my mother would always tell me that whatever I do in life

I should strive to be the best. At this point I changed my attitude and buckled down. I wanted to be a survivor and I did not mind the psychological challenge of what the army had to offer. I actually looked forward to physical training because it helped build endurance and allowed me to become self confident. We marched miles upon miles, did push ups, chin ups, participated in grass drills and trained on the monkey bars. We also became familiar with light arms weapons such as rifles, pistols, machine guns, hand grenades, rocket launchers and trained in hand to hand combat.

My platoon began to work as a team. Our drill instructors encouraged physical and mental competition between platoons and squads. Individual competitions were held to look at a persons performance. A person who excelled was given special privileges such as being able to leave an area early if a person wanted to go somewhere. Another privilege consisted of getting out of a duty called policing an area and allowed to do something else. Policing an area required a company to pick up stray debris. Debris could be anything from cigarette butts to paper. I was highly competitive by nature and felt that even though I might not like it here, I might as well make things a little easier on myself. The special privileges I received became my personal goal to cope with the environment surrounding me. I was very determined to put my mind and physical attributes to work for these privileges.

Basic 1970

The purpose of basic training was to become physically fit, accustomed to using military weapons properly in combat, escape and evasion if ever captured or separated from your company in a war zone. We were not allowed to leave the base during our basic training. We were given the opportunity to go to the base theater, church or military commissary as long as we were back by 10:00pm. Ten o'clock was the military rule for lights out. On certain nights, it was hard for us to make it back to the barracks by lights out when training extended past 8pm. If a soldier was out and caught past lights out they could receive an article 15. An article 15 was considered a misdemeanor in the service. If a soldier received more than three article 15s, that person could get a court-martial with possible jail time or receive a dishonorable discharge. The consequences of going to jail or receiving a dishonorable discharge is what helped keep me in line. A dishonorable discharge is a very serious mark against a persons character and it stays on record for life. This can hinder a person for getting work, establishing credit and receiving promotions.

Soldiers were well taken care of throughout basic training. If a soldier became sick, the Army had many types of medicines to administer to counteract any types of infection. All potential soldiers were educated about how getting sick in a real war would endanger their lives. Soldiers were made to realize if a man became ill, it would take one person and many others out of their combat duty roles. Getting sick would put a unit at a strong disadvantage because the enemy could strategically overpower a military base. The likelihood of a soldier getting sick was slim because they ate nutritious meals, kept their system up to speed and the overall condition of their body in top shape.

I was talking with the men in my barrack when I heard a person walk up to the door and yell, "Private Gregory!" The loud drone of the Sergeants voice made the talking in the room completely cease. I looked at the clock and I saw it was round 7pm. I thought it was rather unusual for a Sergeant to be summoning me for God knows what at this time of the evening. I walked over to the Sergeant and discovered the Captain had wanted to speak to me. The Sergeant escorted me to HQ (Company

Headquarters) and directly into the Captains office. The Captain excused the Sergeant the moment I walked into the room. The Captain began by asking me why I turned down West Point, OCS, WO, ASA and NCO schools. I told the Captain I did not want to make a career in the service and I felt the military lifestyle was not for me.

The Captain was of a stern nature but remained pleasant throughout our conversation. He told me he would give me one day to think about the offers. The Captain advised me to keep an open mind and to consider the possibilities of how I would benefit from these opportunities. The Captain said if nothing else I should seriously consider enlisting in NCO school. Non-Commissioned Officer School would require less years spent in the military then the other four options. He told me if I accepted the NCO offer, my three years would go by very quickly. The Captain promised me when I adjusted to the military lifestyle I would be allowed to transfer to Officers Candidate School at any time. He warned me about turning down West Point. I realized if I turned down West Point I would never be offered such an exclusive opportunity again. The Captain told me he did not want an answer from me right away. I was ordered by the Captain to give him an answer by 6pm tomorrow evening. This Captain was the Captain on duty with the Sergeant at night. This Sergeant was not part of my drill instructors team.

The next day arrived and I did not feel like talking to the Captain or his Sergeant. I did not want to give him my answer because I knew I would be lectured and forced to hear a speech. I wanted to avoid any extra harassment on my part. I never went to see the Captain at six o'clock in the evening the day I was supposed to. At exactly 6:10pm the Sergeant walked into the barrack I was in and yelled "Private Gregory!" I was relaxing in my bed when the Sergeants bellowing made me jump up and stand to attention. The Sergeant asked, "why I did not report and if I understood what 1800 meant. It is 6:15pm and you are late." Once again the Sergeant escorted me back to company headquarters where he ordered me to walk in and see the Captain. I removed my cap before entering the Captains room. I walked up to the Captains desk where I stood at attention. I looked at the Captain and said, "Private Gregory reporting as ordered sir." Before the Captain put me at ease he asked me if I understood what eighteen hundred hours meant. He asked me this question because I was over twenty minutes late. I told the Captain I understood without making any excuses. The Captain looked at me and said, "at ease Private Gregory." I put my hands behind my back and spread my legs in a relaxed manner.

The Captain asked me if I had an answer for him about the opportunities we discussed yesterday. I told him I arrived at a decision. I told the Captain I was not going to accept any of the offers because I did not want to remain in the service past my two year limit and I did not want to make a career in the military. The Captain told me I will never have any of these career opportunities offered to me again and I should consider these opportunities a little more carefully. The more I rejected the Captains logical reasons for accepting these opportunities, the more irate he became. The Captain and I clashed for over an hour that evening. The Captain looked at me and said there is nothing more I can do and excused me to tell the Sergeant to come back into his office. I told the Sergeant the Captains message and he told me to wait outside.

I was getting tired of waiting outside for the Sergeant to give me permission to leave. Twenty minutes had gone by and still no sign of the Sergeant. I walked back to my barrack to kill some time. The front of my barrack was within seeing distance of the headquarter door. I talked to one of my friends for ten minutes. I left my friends when I saw the Sergeant walk out of the headquarter door and stand on the wooden porch. The Sergeant scoped the area to see where I had disappeared. The Sergeant spotted me walking from my barrack. He looked in my general direction and yelled "Private Gregory get over here!" I could hear my friends making comments in the distance. They wondered what I did to make the Sergeant so angry. I ignored my friends and I did not turn around to look at them or make a statement back. It would have been a death sentence to ignore a Sergeant in such a critical situation. It seemed like it took me forever to walk up to the Sergeant because I was so scared. Sergeants had the gift of putting the fear of God into a person.

The Sergeant waited for me to walk up to him and look him directly in the eye before he spoke to me. Then he took a few steps closer to me and began talking in my face. The Sergeant asked me angrily, "what is wrong with you? Why didn't you accept any of the Captains offers? Private Gregory, I am going to give you one last chance to reconsider. I want you to go back to your barrack and think about the answer you will give me."

The next afternoon the Sergeant approached me during our formation. "Private Gregory, have you reached a decision?" I replied, "Yes, I have Sergeant. My answer still remains the same." The Sergeant nodded sternly and told me tomorrow morning I had latrine duty. Latrine duty consisted of cleaning the military bathrooms. There were brass fixtures that needed to be polished, commodes and sinks that require cleaning, floors to be mopped, showers that needed maintenance and stains that needed to be removed from the urinals. Latrine duty lasted for seven days. This particular duty went two days beyond the normal rotation of a regular duty. Latrine duty would have gone beyond the normal rotation cycle had I not have been released. The reason I was relieved of latrine duty was because I was scheduled for kitchen patrol. Kitchen patrol consisted of general kitchen duties. Each person is assigned a specific job and once they performed that job you would be allowed to switch to a different job. There was various busboy duties that needed to be performed. Certain duties required knowing the amount of people in your unit so you would know how much food to prepare for distribution. Another type of job included serving people food on a tray. There were pots, dishes and pans that needed to be washed at the ending of each meal. Tables required cleaning and floors needed to be mopped after each dining session. My kitchen duties were supposed to last five days but for some reason they were extended to eleven days. On the eleventh day with no relief in sight a Colonel happened to walk in. He said hi to me and was talking to the head cook. As they were walking out of the utility room where I worked cleaning pots and pans and peeling 250 potatoes a day. I heard the Colonel say to the cook, "Hold on a minute." He walked back in by me and said, "did I see you a couple of weeks ago doing latrine duty and saw you here the week after." I said, "yes sir." He then asked, "how long have you been doing kitchen duty?" I told him, "I am on my eleventh day." He then asked, "So I get this straight, you went from latrine duty straight to kitchen duty?" I again said, "yes sir." He then asked, "how long was I on latrine duty?" I told him seven days. Right there and then he turned to the head cook and started to yell at him and told me to leave. He told the cook I was missing important training and he better never find out anything like this happens again. I did not mind the extended duties because it felt good to escape running the first couple of miles early in the morning and performing vigorous physical training exercises.

Drill Sergeant Brown approached me and mentioned some officials were discussing why I had turned down all the exclusive career opportunities the Captain had offered me. Even Sergeant Brown was asked to intervene to try and do something to make me change my mind. Sergeant Brown remained very pleasant during our brief conversation. He explained to me I would be receiving the best possible education from five of the finest schools. He explained the tremendous impact that graduating from an honorable school would have on my life. The Sergeant elaborated on the necessary skills people received from these schools to help them be in demand in the working world. The Sergeant mentioned the money was not bad compared to the salaries people were making on the outside. I told the Sergeant I was making more on the outside before I went in to the service. In 1970 receiving one hundred and sixty dollars a day was not doing badly. The Sergeant looked at me and said I was one of the fortunate individuals who was doing well on the outside. The Sergeants last efforts was to tell me that I should not judge or envision the army by what I experienced in basic training. I told the Sergeant that I have not changed my mind and declined the five offers. Sergeant Brown said, "OK Private Gregory I hope you realize what you are doing." This was the last time I heard anything mentioned about the five offers of military schooling. After all of the lecturing I received I was finally allowed to get back in the normal routine of the company.

Basic training lasted two months for our company. On October 2, 1970 each platoon was given their orders for AIT (Advanced Individual Training). Ninety percent of the guys went to infantry. The other ten were given other duties. I hope and prayed I could have been one out of the ten percent who avoided infantry duty because I knew I would be going to Vietnam. As the list to place soldiers became shorter they finally called my name. I was instructed to report for Advanced Individual Training with Company D 3rd Battalion 3rd Brigade. This particular company I was assigned to was an infantry unit. The reason I was placed with an advanced company was due to the high potential the Sergeant saw in me and my outstanding drill scores. Company D 3rd Battalion 3rd Brigade was so unusual because the people placed here had at least one year of college education. I was now assured military duty in Vietnam.

After everyone received their orders, we were instructed to line up for formation. We received a speech, saluted and were escorted in a march by our new drill instructors to our advanced individual training company immediately. We were introduced to our new drill instructors, assigned a new barrack, platoon and squad. My barrack had quite a few new faces. The drill sergeant allowed us to relax for the remainder of the day. The first thing I did was settle into my barrack and organize all my personal belongings into their rightful place. I familiarized myself with the barrack and noticed a difference in the way things were structurally laid out. This particular barrack was arranged differently than the other barrack I lived in. Individual privacy came with some of the changes. We had more room between our beds in the sleeping quarters. The lavatory consisted of more showers and sink space and the commodes were finally partitioned off. This resulted in avoiding the normal procedure of standing in line just to do one thing. I noticed the barrack was a lot cleaner than the first one. One thing that made life a little easier was the recreation area we were entitled to have. Our recreation area had one pool table and a ping pong table. These new perks helped me get my mind off of things. I became a very proficient pool and ping pong player. I was surprised that a lot of other guys did not take advantage of these privileges for a pastime. Instead some would leave the barrack and drink alcohol until they became drunk.

The army was ready to begin building us up mentally the way they wanted us to become. At this point, we were allowed to use our minds and everything we did focused on mental ability. Our drill instructors preached that if we were going to make a mistake it had better been during training then in Vietnam. If a mistake was made in Vietnam it could mean death to you or your fellow soldiers. In Vietnam there were two classifications of people, the quick and the dead. We discovered that the most important lesson in life was to learn by our mistakes and to be able to pinpoint where you went wrong.

CHAPTER 3

No Looking Back

On October 3, 1970 our drill instructor entered our barrack at four thirty in the morning and screamed at us to get up. He thought we weren't getting up fast enough so he began to bang his wooden black stick on each individual medal bed post. He walked through the room yelling that we had better be lined up in formation outside in ten minutes. Everyone knew at that point things were going to start getting serious. Their job was to fine tune the skills we learned in basic training. We were warned if one person screwed up the entire company would suffer. In Advanced Individual Training we were given the opportunity to obtain weekend passes. The passes allowed us to leave the army base from Friday night 6pm to Sunday night 10pm. This would go into effect in our third week of advanced training. Our performance as a company was low and not up to speed. We were warned if we wanted to get our passes we had better improve by the following week.

Our company concentrated on physical fitness and put more emphasis on striving for perfection in our expertise of weapons and knowledge of hand to hand combat. I was at a point where I wanted to become more proficient. I enjoyed hand to hand combat because it taught me the art of mastering the strategies of Jiu Jitsu. One day after an intense hand to hand training exercise our drill instructor said, "you have been trained in the art of Jiu Jitsu and for those of you who have been street fighters, if you combine what we have taught you in Martial Arts with your skill of street fighting, no one can defeat you."

A.I.T. 1970

I looked forward to shooting high powered weapons such as the M-16 rifle, M-60 machine gun, LAW (Light Anti-Tank Weapon) and the M-79 grenade launcher. We learned how to properly throw grenades and became knowledgeable about the use and handling of plastic explosives. We practiced using military weapons and learned the techniques of self defense and survival in the jungle. During our training we were taught to kill as well as to maim. Most of the military weaponry was made to maim a person. Our instructors told us constantly If you wounded an enemy, you would be taking more than one person out of action. It would require two people to carry that enemy back and possibly a vehicle. There would also be a doctor and his staff involved to take care of him. The psychological ramifications of maiming are extremely helpful in fighting a war. They make people realize the outcome maiming has on a person. Once a person is wounded in a war they may be able to return to the front line to fight again. If they can't return because of a serious injury they would be sent home immediately to their families. When a wounded person is sent home from a war because they cannot function normally, fathers and mothers become less tolerant and may speak out against the war. Children would be discouraged from enlisting in the services. We realized that war is not won by eliminating the enemy but by wearing the enemy and their supplies down.

Our marches became longer and faster and extended twenty miles at a time. We were lucky the temperature in Fort Lewis, Washington was cool when we had to march such long distances. The weather was different in Washington than any other state. I had never seen so much gloom and rain as I did in the state of Washington. The last month of my training it rained during the daytime and cleared up every night. During our speed marches a lot of guys would fall to the ground because they were out of shape and could not handle the fast pace of walking the long distance. One day I left the speed march formation and walked over to a man lying on the ground. I watched the man take his helmet off as steam rose from his head. I asked him if he was alright but I could not hear his response because the Drill Sergeant began screaming at me. The tone of the Sergeants voice was very indignant as he yelled at me to leave the guy alone and let him remain on the ground. Then he reprimanded me for abandoning the company and stopping to help one man. The Sergeant told me this man lying here on the ground is a dead man. He said If this was a real war situation and I had left my company, I would not only have endangered my life but the lives of my fellow soldiers. The Sergeant told me to catch up to the platoon. As I left the guy laying on the ground, I could hear the Drill Sergeant tell him that he was a dead man and endangering the lives of the men in his unit. I thought about the man lying on the ground and hoped no one would leave me behind in a hostile territory. The purpose of a speed march was to cover a lot of territory and gain time by walking fast. When we marched we were dressed in combat gear, carried thirty five pounds of military equipment in our back packs, wore a helmet and held a rifle in our hand.

Our marches took place in a hilly and wooded environment. I used to look forward to training in such a majestic place because the landscapes were so breathtaking. Pine trees filled the area with a fresh aroma and they looked like huge towers reaching for the sky. Everything in this particular forest seemed to have grown beyond its full capacity. Everywhere you looked you could see different displays of trees in shades of green. Even the ferns had an unusual look. I had never seen a fern grow to be seven feet tall and have leaves over fifteen inches wide. The overall area had a prehistoric setting as steam settled over the horizon at the beginning of the day. A person had to be careful walking in this forest because of the six foot ant hills and six inch wide ant trails. The farther you walked into the forest the denser the foliage became and smelled musty due to decaying vegetation.

One cool morning our company marched to a very beautiful area. We stopped because our First Sergeant wanted to explain some details about what we were supposed to do in this training session. After the First Sergeant was finished speaking he called the company to attention. I did not hear the First Sergeants command to stand at attention because I was engrossed in the beautiful scenery. The First Sergeant called my name when he spotted me looking around and not paying attention. I did not hear him the first time. The First Sergeant called my name a second time. I saw everyone standing to attention so I snapped to attention. The First Sergeant asked me if I was enjoying myself. I was startled and realized I heard most of his discussion but I tuned out the last five minutes of his speech. The First Sergeant told me to stand in front of the company and knock out ten push ups. I performed my push ups in silence and then stood up. The First Sergeant said he could not hear me and told me to knock out ten more. This time I yelled counting one, two, three and so on to ten as I performed my push ups. The First Sergeant said he heard me but I had to get back down and do them right. I got back down and yelled one Drill Sergeant, two Drill Sergeant and so on. I jumped back up. The First Sergeant corrected me by saying, "I am not a Drill Sergeant. Give me ten more," he ordered, "and this time do them right." I was fortunate that a Platoon Sergeant was right there and told me that he was a First Sergeant and I did my ten push ups properly finally. This was beginning to get old I thought and not to mention uncomfortable because of the heavy gear I carried on my back.

CHAPTER 4

On to Vietnam

I left infantry training at Fort Lewis, Washington on December 5, 1970. I was headed home for a 24 day leave and had to return December 29, 1970. I enjoyed my leave the best I could knowing my next stop is Vietnam. Christmas was not the same not knowing if I would be here for the next Christmas. I did enjoy the home cooked meals.

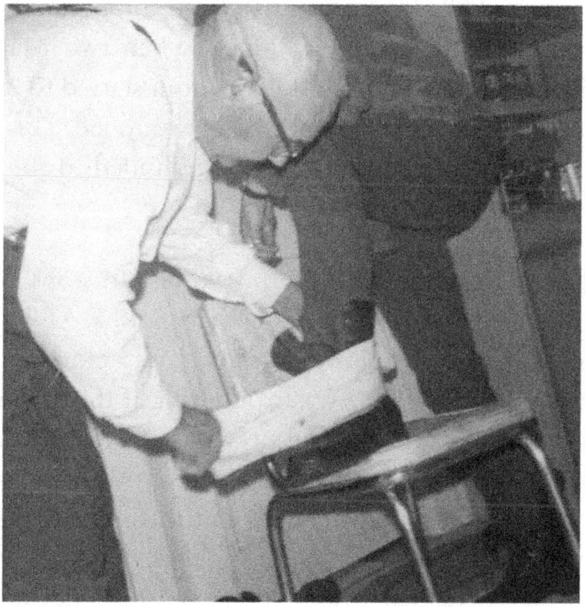

Grandfather polishing my boots the night before leaving

On December 28, 1970 sadness seemed to overcome me, my mom and my girlfriend. I did my best to talk about everything but Vietnam. Then when it was time to board the 747 at O'Hare in Chicago, I gave my mom a hug and told her I loved her. Then I gave my girlfriend a hug and a kiss and told her I loved her. They both had tears and at that point I could not hold back any longer. I turned and said I have to get going. I wiped my tears and headed onto the plane.

I spent the night in a holding company barrack with some of the guys I went through A.I.T. (Advanced Individual Training) with. The next morning we turned in our dress uniform and stateside fatigues. We were then issued our combat jungle fatigues. Once we got back to the holding company the guys started to talk. Vancouver British Columbia was not that far away. If the guys made it there, Canada did not have to arrest or extradite draft dodgers or military personnel. It ran through many minds. I figured 1 year away was better then exile the rest of your life.

The morning of December 30, 1970 we boarded a Flying Tiger Boeing 707. I felt like this was the beginning of the end. I had three cousins go over to Vietnam and all 3 were wounded. My mom and I were over at her sisters house in Calumet Park, Illinois when two Marines in dress uniform and a priest came to the door. My aunt started to cry when they told her, John, her oldest son was seriously wounded. This is why there was an uncertainty to my survival. Three wounded in the family made my odds pretty slim.

We took off from Seattle, Washington making two stops before landing in Vietnam. Our first stop was in Anchorage, Alaska. As we were on approach you could see snow plows on the runway plowing ahead of our landing. They looked so close I was wondering if they would get out of the way in time. When we touched down, the runway must have been icy. The plane fishtailed as we went down the runway. The pilot throttled the left and right engines to gain control of the aircraft and kept it from sliding off the runway.

A Sergeant escorted us into the terminal where we were brought to a large room and told not to leave this area. The Sergeant did tell us it was going to be about 4 hours before we leave. We had to wait for the snow to subside. At this point I did not care so I left the room. I found a bar in the terminal and sat down. The bartender asked where I was going. I told him Vietnam. He said then I won't ask for an I.D. and the first drink was on the house. I finished that beer and ordered another. He then asked for $4.50. Now this is 1970 and beer back home was 50 cents. The gentleman next to me bought that beer and wished me luck in Vietnam. Two beers were my limit so I headed back. As I walked back into the room where the troops were at the Sergeant yelled, "where is Gregory?" I said, "right here Sergeant." He started to ask me where I was at and then said, "never mind I should have joined you."

After the four hour delay, the weather cleared and we took off headed to Tokyo, Japan. This was the longest of our flights. We landed at Heneda Airport which was on the outskirts of Tokyo. Again our Sergeant escorted us to a holding area and said we have about two hours before we leave and do not leave the area and that goes for you too Gregory. As soon as the Sergeant left, I heard a couple of guys saying they needed a cigarette. So I said we have a couple of hours lets go outside. Three of four guys came, one got cold feet and went back. We went outside. It was absolutely beautiful. By the appearance of the Japanese people maybe this is what we were to expect in Vietnam. The guys asked are we going to get in trouble. I said, "what are they going to do to us?, send us to Vietnam?" They said yeah really what can they do to us. We went back in and the Sergeant didn't say a word to the two with me but kind of yelled at me with a smile and said, "I told you not to leave this area." I told him I had to go to the bathroom. He said,"all this time?" I replied, "yeah." No more was said.

We took off from Heneda Airport headed to Cam Ranh Bay, Vietnam. We landed at about 6:30am January 1, 1971. What a Hellhole Happy New Year. You get off a plane where its 70 degrees to a temperature in the nineties with high humidity that takes your breath away immediately. Then there is the smell of burning shit, aviation fuel and diesel fuel. Sand everywhere. The wind would blow it and it would get in your eyes, ears and mouth. Then there are the dust devils. I had never seen one so I got close and got sand blasted. They are like a miniature tornado. The Sergeant escorted us to our holding company where we could shower and sleep. The next morning we got up early and were given our assigned duty.

CHAPTER 5

Being a Grunt

I was attached to HHT 1st Squadron 9th Calvary. On January 2, 1971 about five of us were taken to Bien Hoa by helicopter for additional infantry training. We learned about booby traps, how to use a flame thrower, gorilla tactics and the critters that can kill you in the jungle. There were tigers, elephants, cobras and the crate. The enemy would use poisonous snakes in booby traps. We were warned not to kill water buffalo. The Vietnamese used water buffalo for transportation and farming. This training lasted 20 days then I was the only one transported by helicopter to my assigned unit in Phuoc Vinh.

I arrived at my new unit D trp 1st Squadron 9th Calvary 1st Air Calvary Division on January 21, 1971. I was assigned a cot and a foot locker in our living quarters called hooches. Hooches were big open rooms with a privacy area in the back for Sergeants. These were made of wood with solid wood half way up the walls then screen the rest of the way up. The screens were covered on the exterior with wooden louvers to keep the rain out.

My first night here we had a rocket attack. Phuoc Vinh is a larger Army Base consisting of Infantry, Artillery, Air Calvary and all supporting units. A siren went off and guys started running. I asked whats going on. Someone yelled we have incoming so I followed. We went in a bunker between the hooches fortified with sand bags around the outside and 3 layers on top. Nothing was happening so I went to the entrance to see. This rocket a few hundred feet above us was sputtering with sparks out of the back. It flew about a block more and fell into the compound with a massive explosion. After a couple of weeks being with D trp 1/9 the Platoon Sergeant called me into his room in the back of the hooch. He said I hear you do not drink. I told him I really do not want to drink. He said well have one for me. I declined. Then he said get the fuck out of his sight. I don't trust you. I thought here we go. Well evidently he was not well liked for whatever reason and was fragged while he slept. I have no idea if he lived or died and at that point I did not care.

I found out in a hurry nobody wanted to be around the new guy. Our Platoon Sergeant had to assign a guy to help me get through the first few weeks. I was given the M-16 and 21 magazines which were mine for my time with D trp 1/9. I was issued a drivers license for up to a duce and a half, a meal card and a ration card. I was briefed on what we do and after that I was on my own.

D trp 1/9 operated check points, escorted convoys, rescue missions, re-con, ambushes, secure downed helicopters, pulled security for demolition teams and Special OPS (Special Operational Forces) and assess damage from artillery strikes.

D trp 1/9 worked as a rifle platoon for A trp 1/9. A trp 1/9, a Scout unit using OH-6 LOH (Light Observation Helicopter) nicknamed the Loach to recon areas of enemy concentrations and bunker complexes. This is where D trp 1/9 comes into play. Our rifle platoon was 25 men strong. We were called to be airlifted to a bunker complex to see what was there. We loaded four UH-1 Huey Helicopters with six fully armed riflemen per helicopter. Typically we had two men carry an M-79 grenade launcher, 2 M-60 machine gunners, 2 radio operators and the rest with M-16s with at least 21 magazines per person. Everyone had 2 to 5 fragmentation grenades except for the machine gunners and everyone strapped two 100 round belts of M-60 ammo around their neck cross bodied.

The ride in the helicopter was cool. The bad part is a lot goes through your mind when you are on your way to the LZ (Landing Zone). You start to think is this my day to die. Then when you are dropped off the chopper it does not touch down. The grass is about three to four feet tall so the chopper hovers above the top of the grass and you have to jump off. You do not know what you are jumping into. Every jump I made was into the tall grass with 12 to 18 inches of water. The water is loaded with leeches. The guys that did this before did not tell me to tape my pant legs to my boots sealing access to my legs. Well my first time I did get a leech on my leg, I knew something was on my leg but did not know what. I squeezed what I thought was my leg and did not feel me pinching my leg. I grabbed whatever it was and jerked it off my leg. As I held this leech it squirmed wildly and it was about ten inches long and a half inch in diameter. It scared the living hell out of me. I threw it as far as I could. Blood was running down my leg from where it had attached. My arms were also cut from the razor grass. That is the other thing they didn't tell me about. I had my sleeves rolled up, everybody else had their sleeves rolled down. Well I learned in a hurry and during this whole time I am looking for enemy movement.

One of the scariest things about being inserted into an LZ (Landing Zone) is that the helicopter has to drop you off in a clearing. We have to get off and walk to the tree line of the jungle. Between us and the jungle are poisonous snakes, leeches, spiders and who knows what else. Once you make it to the tree line you are on high alert. A sense of false security comes over you. Then you realize these shrubs and thin plants can't stop bullets. You are tense the entire time until you are picked up and headed to base. Then the adrenaline starts to leave your body. You feel exhausted because the entire time you are looking for booby traps, tripwires and the enemy.

On February 7, 1971 we were called out to secure a five ton truck that was ambushed a couple of hours before sunset. We scrambled to help if there were any survivors. When we arrived we found the truck on fire and 5 dead. We formed a perimeter to protect ourselves and the medic at the ambush site. The wrecker came about 2 hours after and now the sunlight is getting dim. Our Sergeant became aware of the sun setting and put an urgency on getting us out of there. The wrecker driver wanted to make sure the fire would not start up again before hooking up to it. We got back to the gates at sunset. Even outside the gates there is a sense of insecurity.

Everybody that starts out on gun jeeps starts as a gunner standing behind an M-60 machine gun sitting on a pedestal in the center of the jeep. It was exactly like the T.V. series Rat Patrol. We would recon our perimeter and 105mm artillery strikes near our base. In 1971 the radar at Phouc Vinh could pick up a mortar or rocket ten feet off the ground when fired. That information is sent to the 105mm Battery and they could usually get rounds off before the mortar or rocket impacted and still have time to head to a bunker. One time after a mortar attack and the 105 Battery fired at the presumed mortar sight, we were called out to assess the 105 strike. The only thing we found was a mortar tube that was bent and had shrapnel holes. There were no enemy personnel present but there was blood on the ground.

We escorted convoys that had tankers and box trucks with our gun jeeps that carried crucial supplies such as aviation fuel, diesel fuel, gas, food, ammo and medicine. By this time I had become a driver and drove point with our Sergeant and Cambodian Scout. Cambodian Scouts were usually defectors from the NVA (North Vietnamese Army). They knew trail markers by the enemy warning if there are booby traps or mines in the area. One day leading the convoy Tuc Noon our Cambodian Scout tapped me on the shoulder and pointed, then the Sergeant yelled, "STOP!" It was for a good reason, the VC (Viet Cong) put up a tripwire across the road attached to a 175mm artillery shell. Our Sergeant called out our demolition specialist to deactivate the explosive. It took them a couple of hours to get to us. They deactivated the explosive within minutes. We then were allowed to pass and after we were a safe distance away, the demolition team detonated the 175mm explosive round. That incident put us over two hours behind and by the time we got to Song Be we were driving in the dark with no headlights for over an hour. The large trucks in the convoy were governed out at 45 mph that slowed our progress.

On February 28, 1971 we stayed at Song Be for three nights. The second night we were there they gave us free fire at our perimeter. Everybody let loose with everything we had. I am really surprised that nobody got hurt. Hand grenades going off, M-79s blasting in the tops of the trees, the sounds of M-16s, M-60s and the clacking of the 50 caliber machine gun. All of a sudden a cease fire order was given. We were told there was a seven man recon team in the area where we were firing. You would have thought someone would have known this before we started firing. Fortunately none of the seven were injured. The same night as I was walking back some guys were crying and others were trying CPR on one of our guys. His nickname was Bosco and he had been in country one month. We were told three days later he had died.

Bosco was airlifted in the morning and our mission to escort the convoy back to Phuoc Vinh continued. This time I had a trailer attached to my jeep. The Sergeant and Tuc Noon got on to my jeep and we proceeded to the front of the convoy. We were about an hour into our drive back and the front universal joint broke and the driveshaft fell to the ground, picking the jeep up spinning us almost 90 degrees. The convoy stopped at this point, then the Sergeant gave the order for the next gun jeep to take point to lead the convoy. We would have to wait for a wrecker to get us. I actually did not realize how big this convoy was until they past us, leaving us to wait for the wrecker. Our gunner Tuc Noon jumped on the jeep taking Boscos place. The Sergeant kept telling me not to fire unless I was sure of enemy presence. He did not want to give our position away to possible enemy forces in the area. It took a long 40 minutes for the wrecker to get to us. He hooked us up and we were on our way. The convoy did slow down for us to catch up. When we did join them the Sergeant went to the point jeep and I had to ride on the back of a semi-tractor. The dust was horrendous. I took off my shirt and wet it with the water I had in my canteen and covered my nose and mouth. The problem with that was it dried fast and got under the shirt. I had my goggles and even those were too hot to wear. Then I tried a gas mask the driver had. That was still too hot to wear. When we got back to Phuoc Vinh I was covered from red dust from the road. Then we started to prepare for D trp 1/9 to stand down.

CHAPTER 6

Volunteering For Ground Missions

One day our Lieutenant came walking up and asked for 6 volunteers. Five volunteered right away and one of the guys called out my name, then the Lieutenant said, "OK Gregory you are the 6th." Before I could say anything 2 of the guys said we trust you in the field. That was one hell of a compliment to me. We were ordered to be ready at 1700 with gear on and at the flight line. From there we will get our orders from the waiting Huey helicopter pilot. We were to have 2 radios, M-16s and fragmentation grenades. We got to the waiting helicopter that was already running. The copilot got out and told us we were to pull security for the passenger we are waiting for and he would give us instructions on the ground after insertion. Then we were told to board the aircraft. After about 10 minutes one soldier dressed in camouflage including his boonie hat and his face colored with camouflage paint. He carried a large backpack, a M-16 and a long bolt action rifle with a large scope mounted on it. As he climbed into the chopper he did not look up or around. The center seat at the back wall of the chopper was to be left open for him. He sat down and the chopper began to lift off. As we made our way to the runway the copilot leaned over and handed him a large manila envelope. He removed the seal over where the string holds the envelope closed. I watched him unravel the string and open the envelope. He pulled out several sheets of paper and as he went through them he pulled 2 or 3 out, leaned forward, tapped the copilot on the shoulder and handed him the papers. It looked like one of the papers was a map. The copilot leaned over and handed the pilot one of the papers. At this point I can see this is some scary shit I am involved in. Only this guy and the pilot know where we are going and as scary as this guy looked I was thinking we are going into something real bad. It was quite a long ride. It took a good hour. Finally we went low level skimming the top of the trees. When we came to a clearing the chopper touched down. We all got off and the chopper took off. We followed the scary guy to the tree line. He stops and puts up his fist and waves his arm in a gesture telling us to get down. Pointed at the radio operator to come forward to him. He gets on the radio and as soon as he gets off, he tells the radio operator the chopper will be back after refueling and we are to wait here. With that he got up and went into the jungle never to be seen by us again. Now there are 6 of us with M-16s in a clearing with no heavy weapons like a M-60. I still think about how scary he was to this day and who he may have been and what was his mission.

My radio with machete foreground

I was with D trp 1/9 for about 4 weeks and one day I walked into our hooch and here's Nick laying on his cot throwing his bayonet and sticking it in the wood floor. I said, "hey Nick what are you doing?" He said, "there's a rat that comes up through that knot hole and I am trying to stick him." He then said, "we have to keep quiet and if you are going to watch you have to move over there." He was setting small pieces of cheese about an inch from the hole. The rat would stick his nose up and sniff the air. It obviously knew what Nick was doing. It would come up fast and grab the cheese and in a flash it was gone back down the hole. Nick did not stand a chance of getting him this time. Then here comes the rat again. This time the cheese was stuck to the floor, the rat hesitates to make a second grab. Nick throws his bayonet just as the rat is going down and misses next to the hole. Nick says damn, gets up and walks out of the hooch. Nick was a strange person. I really did not know what his function was in our unit. He never went out with us on missions.

One of the most scariest times was on February 7, 1971 when I volunteered for an ambush mission which consisted of a ten man team. In this team, we had two radio operators, two M-60 machine gunners and the rest with M-16s, grenades, claymores and a shit load of ammo. Anytime we went out I wanted either the M-60 or the radio. In my head if I had the radio and I was the only one left I could get myself out. With the radio I have the map, the strobe light, machete and compass. I also thought if I had the M-60 it had all the lethal firepower I needed. You could not hide from a rapid fire 7.62 machine gun round. Even a 12 inch tree could not save you.

On this ambush mission we had a green 2nd Lieutenant and a seasoned Staff Sergeant with us. The 2nd Lieutenant decided not to follow the pre-planned route. While we were to follow the tree line to our designated ambush sight, he decides to take us into the jungle. A couple of hours pass as we trudged through thick foliage. The foliage was so dense you could hardly see the guy in front of you. You had to catch a glimpse of his head, shoulder and sometimes the brush would be closing in front of you. All of a sudden the guy stops in front of me and hands me a bottle of salt tablets. This was the first time out of all the times I have gone out in the bush. So I waited for the two guys behind me. I handed the salt container over to the guys behind me. I then turned around and no one was there. I started walking and didn't see anyone after walking through the brush. No broken twigs or any other signs that someone had walked through there. At that point I said to myself, "Oh Shit! They left me!" I waited for the guys that were behind me to catch up but they never did. The jerks did a 90° turn right where I was handed the salt container. I headed back to where I thought I had lost them. Dusk was starting to set in. I figured at this point that I was going to have to stay the night. I began to click my radio mic because we were on radio silence but there was no response back. Every time I went out I replaced my batteries in my radio and strobe light. I am at this point ready to bunker down for the night and will call for help in the morning. Still no response from the forward radio operator. 45 minutes to an hour later they came back for me. The forward radio operators radio was dead. The 2nd Lieutenant and the Staff Sergeant got into a heated argument and the Sergeant took charge of the mission. The 2nd Lieutenant got us lost and not near our ambush destination. The Sergeant called for an extraction using my radio.

Going out on a ten man ambush team

A trp 1/9 sent a Huey to extract us. When we heard the chopper we were to use a strobe light. The 2nd Lieutenant used his radio operators strobe light and that was dead also. The Sergeant grabbed me for my strobe light and by that time the chopper past us. It took another 15 minutes guiding him in by sound because of how dark it was by this time. As he closed in on us the Sergeant turned on the strobe light and the chopper finally sees us. As we are boarding the pilot said we can only take eight and we will be back for the other two. I was one of the last to get on and said I am not staying. So the pilot said get rid of some gear. We shed ammo, claymores and grenades. The pilot may have been red lining but too bad. The way I felt if one stays we all stay. It is bad enough they left me in the jungle for over 45 minutes. They realized they made a 90 and came back to tell me but by that time, I already went straight and they would not have come back if it were not for the other radio being dead.

One of the worst missions I volunteered for was with a five man demolition team. We had 2 M-60 gunners, 1 radio man with an M-16 and 2 demolition experts from another unit. We were strictly security for these 2 men. We boarded the Huey at approximately 0800 hours and took about a one hour ride to our drop off. Again the grass was about 3 to 4 feet tall and the chopper could not set down so we had to jump off. I have a 100 round belt locked and loaded in my M-60 machine gun, another 200 rounds around my waist and another 200 rounds across my shoulders and chest.

Going out on 5 man demolition team

That is a lot of weight jumping out of a helicopter three feet above the ground. Once on the ground the chopper takes off and the demolition experts get their bearings. We are still in the clearing, not heading to the tree line. The demolition guys said follow us. We headed right down the middle of this clearing. This is something you do not do as an infantryman. You always seek cover. We followed them for about 2 miles walking through the tall grass. We came across animal trails by the way the grass was laid over. We then crossed a high speed trail. The grass was gone with nothing but dirt on this trail. It was at least 8 feet wide leading from one jungle to another with an opening that looked like a tunnel entering a mountain. The demolition guy said after passing this high speed trail to wait here and if we are not back in one hour to call for a taxi, meaning a helicopter ride out. The demolition guys went off and disappeared. I told the other two guys with me that we are too close to this trail. We then agreed to move in the direction the demolition team went. Then we hunkered down and waited. We stayed below the grass height not to expose ourselves and keep a 20 foot distance between us looking up every once in a while to see if they were coming back. Fear came over us not knowing what is going on. Then about 45 minutes later they were running and hopping through the tall grass right toward us. We all got into a combat stance thinking there had to be a lot of bad guys chasing them so we prepared for the worst. As they got closer, they started to wave their arms and yelling "RUN!" By the time we turned around they passed us. We ran after them trying to keep up because we were loaded down with a radio and ammo. We ran a good five to ten minutes, then all of a sudden, they turned and said, "GET DOWN!" Then the biggest and loudest explosion happened that I ever experienced other than the incoming rocket at Phuoc Vinh. The shock wave flattened the tall grass as it passed over us. Calm as could be, one of the demolition guys got on the radio and called for us to be extracted. One of the demolition guys told us to be alert because every bad guy within a couple miles know we are here. They had us form a circle with each man looking in a different direction until the chopper arrived. When we got back from the demolition mission the three of us all said the same thing on how scared we were. To this day I do not know what unit they were from or what they blew up.

CHAPTER 7

D trp 1/9 Stand Down

I had heard a rumor that the 1st CAV was standing down and we would be going to different units. I did like the ride in the helicopter, I just did not like that they left me in the jungle. So I went up to the 2nd Lieutenant who was waiting for us to thank us for volunteering for this crucial mission. Right after I asked him, as a OH-6 (Loach) was passing over head, "how do you get to fly on one of those?"The 2nd Lieutenant said without hesitation, "see that hooch over there? Go there and tell them you want to join up with them."The 2nd Lieutenant asked me to hold off because D trp 1/9 was standing down and equipment has to be taken to Bien Hoa. We got there early in the morning of March 12th. Once we turned our jeeps in we were told to meet back at this spot at 0800 hours and then we would be picked up and transported to the flight line. The three of us asked one of the guys that we were turning our jeeps over to, "what is there to do here?" He told us there was a movie theater and different places to eat. While we were on our way to get something to eat, we saw a large PX and stopped there first. When we walked into the PX, it was overwhelming - way too much to look at! I stayed near the door where the electronics were at and bought a 35mm Yashica camera. The other 2 bought some clothes and we headed out. The first place we came across to eat was La Casa Grande, a Mexican restaurant right in the heart of a war zone. The tacos were excellent with fresh hot cherry peppers. I grew up on hot cherry peppers. My Grandfather grew them every year. After eating, we went to the theater and watched a John Wayne movie.

Returning to Phuoc Vinh, our work was cut out for us, getting our radios, rifles and other equipment back to supply. Some guys were already being transferred, so on March 16th I decided to take a walk across the way to A trp 1/9 Company Headquarters to ask to join up with them. I walked in and a Captain, who was the Commanding Officer, was doing paper work behind a desk said, "Can I help you?" I said, "Yes, I would like to join up with your company to fly." He asked me what unit I came from and I told him D trp 1/9. After that no more questions were asked. He then said, "grab your gear, I will tell them you are over here and your paperwork will catch up." The one thing he left out, that I found out later, is how bad they were getting shot up. The Platoon Sergeant told me to find a cot in a hooch he was pointing at. Then for the next 2 days I was given a physical and fitted for my flight suit, gloves and helmet. I learned these type of units wanted soldiers with an MOS of 11B because we were familiar with M-60 machine guns. In this hooch there was a bunker attached to it. It was scary walking by it. It was a dark hole leading to who knew what at that time. Then I kept seeing 2 to 3 guys going in there and music coming out so I went over and looked in. Small Christmas lights lit up the darkness the best it could and the thick smoke of marijuana filled the air. One of the guys asked me if I wanted a hit. I said I have things to do and just wanted to look inside. As I left I was given a warning to keep my mouth shut.

Shortly after I started to fly with A trp 1/9. When you first come to an Aeroscout Unit as A trp 1/9 you fly left seat next to the pilot. You are taught to fly the aircraft in case the pilot is shot.

Me flying left seat

If the enemy is found or we take gun fire, left seat observer is to drop smoke so the Cobra, at 2,500 feet, knows where to fire rockets and 7.62 machine gun fire to give us cover to escape the area.

CHAPTER 8

Aeroscouts

A trp 1/9 and all other Aeroscout units were divided into three groups. AH-1Cobra Gunships were the Red Team, OH-6 LOH (Loachs) low level observation helicopters were White Teams and the Huey UH-1 with rifle platoon were called Blue Teams. When the Cobra and OH-6 teamed up they were called a Pink Team. On missions a Cobra, OH-6 and a Huey would fly together. The OH-6 would fly as low to the ground as possible. The OH-6 is a low level reconnaissance helicopter with contact. We would look for any signs of the enemy such as footprints, grass laid down, bunkers, trees used as smoke stacks from an underground bunker complex, sand pans sunk to avoid detection, mud stirred up in the water, food containers, the ground disturbed where they may have buried supplies and anything a human can leave behind. I would fly two to three missions a day.

Cobra gunship

When a bunker or bunker complex is found the enemy would stay in hiding. They were deathly afraid of us because we had a Cobra flying at 1,500 to 2,500 feet above us ready to dive at a moments notice. If it were too much for our team to handle we could call in the coordinates for artillery or an air strike. Our job as the low bird was to draw fire. This confirms enemy presence. My job was to fire my M-60, drop fragmentation grenades, concussion grenades, white phosphorus grenades or CS gas into the bunker to piss them off so they would take a shot. When finding a bunker complex we had to determine the size of the bunkers and the bunker complex. All this information is being radioed to Command and Control in real time.

On March 31 the entire company of A trp 1/9 left Phuoc Vinh and moved to Tay Ninh. This place is desolate. It was once a major base for the United States Army with a runway and now in ruins. Our accommodations were tents, dirt floor and cots. No mess hall just outside eating. As time went on we gathered pallets to make our floor and some interior walls. We halved the larger rocket boxes lengthwise to make shelving. Our work in the air did not stop nor did anyone want it to. You could see Tay Ninh was an Army base rocketed and mortared over and over again. Revetments where helicopters parked were blown apart. Small craters and the lack of vegetation around them made it obvious where the mortars fell.

Our missions continued in Cambodia, Parrots Beak, Fishhook, Iron Triangle, Hobo Woods and the Delta. We cleared LZs (Landing Zones), went ahead of armored vehicle units and sometimes flew ahead of convoys to make sure they weren't ambushed. One day flying ahead of a convoy we were ordered to maintain altitude and forward air speed. We were only about 30 feet above the road and slowed our forward air speed. We complied, then all of a sudden SPOOKY, an AC-47 now flown by South Vietnamese pilots flew right over the top of us. Then they went down to about 30 feet and started to fire from both sides of the aircraft. Trees and brush started to fall as the mini guns fired away. It was a spectacular sight to behold. Our job was pretty much over at this point. Nothing could have survived the amount of lead it was pumping out. We decided to go out to the right flank where there were some land depressions where the enemy could hide.

By this time I have moved into the back seat as gunner. In the front the pilot sat in the right seat who is a WO (Warrant Officer), Lieutenant or Captain. In the left seat the Oscar (Forward Observer) and in the rear compartment behind the pilot the gunner with a M-60 machine gun sitting on the floor. I am sitting on an armored plate with legs hanging out the door. We would make low and slow right hand turns usually around 7 to 15 knots, but if it is in a low brush area we may hover to blow the grass or brush apart with our rotor wash. It may expose a bunker, tunnel entrance or even the enemy. You can only fly so many days per month, then you have to be down a day or two.

CHAPTER 9

Chup Rubber Plantation

We began to work Cambodia pretty heavily beginning in May. Working more towards villages, rivers and rice patties. There were some jungle areas that we reconned to the north at the Vietnamese/Cambodian border. We would also recon the rubber plantations. Rubber plantations were very numerous in Cambodia. One plantation in particular was called the Chup Rubber Plantation. This one was known to be an enemy sanctuary. On May 7, 1971 our Hunter/Killer team, that consisted of a 3 man crew on an OH-6 LOACH, 2 man crew on an AH-1 Cobra and a 5 man crew on a UH-1 Huey. The low bird OH-6 flew above the tree tops of the plantation. This was the first team out that morning with pilot LT Frank Bengtson, WO1 David Myer and rear seat gunner SP4 Larry Rothel flew into a Battalion of NVA Regulars (North Vietnamese Army) positioned in the Chup Rubber Plantation. From the edge of the plantation the Scout team was just above the trees and in only about 30 feet. Either the NVA saw them coming or heard them and readied themselves for the attack. The amount of fire power and the lack of tree coverage was too overwhelming and the aircraft was shot down and all 3 crew members were killed. A rescue attempt with the chase bird, UH-1 Huey, flying at the same altitude as the Cobra at approximately 1,500 feet started to descend and as it got closer to the crash site there was a lot of automatic weapons fire and had to pull off. The Cobra did not dive to shoot up the area because he did not know if the crew was alive or dead or if they were running in a certain direction. A second attempt was made by the Huey crew. The Cobra flew at low level in an attempt to draw gun fire away from the Huey crew. They also experienced heavy automatic weapons fire. At this point the mission was aborted until we could get more man power and fire power to extract the crew.

On the morning of May 8, 1971 the mission was to recover the bodies of the Scout crew. I was in the lead OH-6, the first to go over the crash site. The first thing we noticed as we approached the plantation are three white sheets laid about 30 feet outside the rubber tree line in the clearing. We made a fast pass above the edge of the plantation tree line. We did not draw fire so we kept moving in with fast passes. No enemy was in sight. Our passes became more routine with slow right hand turns. We worked a large area to make sure it would be safe for the recovery mission. We gave clearance for the Rifle Platoons Hueys to descend. The Rifle Platoon pulled security for the recovery crew. They set up a skirmish line and then began to move into the rubber plantation towards the crash site. Once the area was secured, the recovery team was ordered to descend to the area to pick up the bodies. Once the crew was recovered the Huey left the area to return the bodies to Bien Hoa. The Rifle Platoon continued to the crash site and saw that the OH-6 LOACH had been stripped down. The radios, batteries and all ammo had been removed. Then they were ordered to be extracted and get to the outside of the rubber tree line where the Hueys were descending to pick them up. We continued flying low level to give cover for the extraction for the rifle platoon. Once the rifle platoon was safely on their way, we flew over the top of the rubber trees to assess the area. As we recon the area, we found a tow behind twin 51 cal anti-aircraft gun and truck left behind by the NVA. What baffled all of us is that the NVA had moved the bodies from the crash site to outside the rubber tree line and covered them in white sheets. During my time here in Vietnam I heard a lot of stuff about the NVA and their treatment of U.S. GIs bodies. This gave me a different perspective. The NVA Regulars are warriors not to confuse them with the VC (Viet Cong). This gesture of moving the bodies out of the Plantation and covering with sheets was respect for us Scouts. This to me was warrior to warrior respecting our bravery in the face of death.

After the recovery we started to pull off. I had to remind Captain Knight, my pilot, that there was a truck and anti-aircraft weapon we did not destroy. Captain Knight called up to C & C (Command and Control) to get permission to return. We were granted permission to return. Once we got back to the area I threw fragmentation grenades, incendiary grenades and shot them up with my M-60 machine gun. Then we marked them with smoke and let the Cobra roll in with rockets and minigun blazing. We did not go back to assess the damage. We flew back to base in total silence. Usually we talk on the ride back. The pilot turned the controls over to the left seat and leaned back with his helmet against the top of his seat. We were all experiencing and reflecting the life and death of our comrades. You feel anger, you want revenge, sadness and the worst part is the silence. This is when you numb and detach. The worse for me was I got to know Larry Rothel. I knew his girlfriend and his family through pictures he had. We both talked extensively about our family back home. He wanted to be a rodeo star after he fulfilled his obligation to his military service. His Mom had sent him his lariat that he used in rodeos prior to military service. That hurt me emotionally and mentally. From that day on I did not want to get to know anyone. After this you had to survive days, weeks and missions for me to get to know your name.

Captain Knight was emotionally and visually upset. Captain Knight was the Commanding Officer of the Scout Platoon and has lost a crew of three, through no fault of his own. He went straight to the Company Commanding Officer and requested an airstrike for the Chup Rubber Plantation. We in the Scout unit knew there was a large concentration of NVA Regulars in this plantation. The first request for an airstrike was denied. The reason they gave was they were busy bombing somewhere else. Captain Knights second request was for B-52s out of Thailand to bomb the rubber plantation. This request was also denied because they said they had inclement weather. Now Captain Knight is getting angry so he ordered an infantry unit to go in and destroy the NVA. The call came back that the infantry unit will not go in without tanks. Captain Knights response was, what seems to be the problem, send in tanks. The third request was denied because the tanks are 12 feet wide and the rubber trees are planted 10 feet apart and we are not allowed to destroy the rubber trees. At this point Captain Knight is furious and shuts down operations until he gets more information and a new AO (Area of Operation) before any more get killed in that AO. Captain Knight left our base for a couple of days to find out why we could not destroy the enemy. When he came back he told me that some high ranking politicians had some type of ownership in the Chup Rubber Plantation and we will not be going back to that AO. We could still work rubber plantations with our OH-6 L0ACH and the gunner using a M-60 machine gun and the Cobra because damage would be minimal.

After an intense mission in Cambodia, in fact the last mission of the day, all the helicopters got back to Tay Ninh at the same time. Another gunner named Nichols and I started walking from the flight line towards our tent. As we were walking, he asked me, "Vern, have you been scared yet?" I said, "Yes." He then asked, "So scared you could not get the tip of a toothpick in your ass." I said, "no." Then Nichols said to me, "Then you have not been scared yet." That really made an impact on me. I have felt the adrenaline rush of a gun fight, but I had the majority of the control. I caught the enemy running, trying to hide or they just realized our direction and started to turn before I dropped them. Then two times after he made that statement, it happened. Both times there were multiple enemy firing automatic weapons at us. I have the hammer down on the M-60 machine gun, going back and forth from enemy to enemy and both times it was with pilots that slowed almost to a hover to watch the action. I could not let go of the trigger to push the mic button to tell the pilot to get the hell out of there. If I could have I would have punched the pilot in the back of his helmet. Then ever so slowly it seemed like he would turn the aircraft left so I could still cover our departure until I had no more targets. As we started to head back, at about 40 minutes into the flight, I could feel something going on in my rear end. It was my sphincter going from extremely tight to relaxing to normal. Now I understand what Nick (Nichols) was talking about.

Missions continued along the Cambodian/Vietnamese border, I am now getting 3 missions a day. We are finding more weapons and food supplies in the jungle. These supplies are sometimes buried or in a bunker. We can see if the ground has been disturbed for burial of supplies so what we do is drop fragmentation grenades or concession grenades to expose what is buried. Sometimes we have the cobra roll in and fire some rockets to try and uncover what is buried. In bunkers I throw fragmentation, concussion or white phosphorus grenades to try and blow the top off or make the entrance larger to see inside or get a secondary explosion. Once we confirm the presence of supplies the Blues (Rifle Platoon) are called in. A rifle platoon of approximately 25 soldiers are sent in to evaluate the situation. If it is a large supply of food it will be destroyed on sight and if it is weapons such as rifles or small arms they will be carried out and loaded on to the Hueys. Ammo, mortars and rockets will be destroyed on sight. We are constantly reconning the area to keep the boots on the ground safe from the time they are inserted into the LZ to the time they are extracted.

The more experience I got, the more the pilots trusted me to pull the trigger. Being a Scout door gunner you could not hesitate. We never went into safe zones. Our intelligence comes from captured enemy that are interrogated. The intelligence that comes out is relayed to us and it is our job to confirm their findings and come up with a plan to destroy what we found. We were always given free fire in the areas we were sent to which means if any living thing whether human or animal is within my range, it is my choice if it lives or dies. I made it home so you know what my choice was. Any contact with the enemy was anywhere from 20 to 60 feet and extremely violent. Then comes the silence. I began flying 3 to 5 missions a day. Depending on the authority of the pilot you could get First Light, which is just as the sun is coming up. You recon around your base to see if there has been any enemy activity or enemy presence. Then you fly to the fire support base that you are working out of for that day, fuel up and fly out to the first mission, then third mission, then fifth mission and finally Last Light around our base compound. When you get 5 missions its called bogarting on behalf of the pilot. I am just a passenger looking for the rush.

CHAPTER 10

A trp 1/9 Stand Down

On the morning of June 1st without notice, we were told to pack our gear and head to the flightline. We were told to board a Huey that will take us to Quan Loi and we will receive further orders from there. There was absolutely nothing to do here. I started to jog to stay fit. We were given weed whips and weed killer in a 2 gallon spray pump container. I did not mess with either one. My buddy Randy Kramp lucked out and left a day after we arrived at Quan Loi. The rest of the Scouts were reassigned immediately. Randy went to Bien Hoa and I am sitting here in limbo. I did not have an address for anybody to write to me. I could write but not receive letters. This gave me time to reflect on what I had done with D trp 1/9 and the rush of flying Scouts with A trp 1/9. One day I am walking back from the mess hall at about 1300 and could see a commotion near my hooch. I looked down as I walked and saw bloody foot prints. I asked whats going on. A Sergeant had been fragged in the room behind the hooch. I never did find out who that was. Orders finally came down for me. I was to report to the 238th in Tuy Hoa by June 14th.

One of our gunners was going home. He was to report to Bien Hoa in ten days. He started to come down from his heroin addiction. He knew he could trust me to help him. There were rumors that if you test positive for drugs they would not let you leave Vietnam until you were clean. He asked me to help him in every way. I got him extra blankets when he started shaking. He keep saying he was cold. I would cover for him when an Officer or Sergeant came in. They would ask me what was wrong with him and I would tell him he has a slight fever and upset stomach. They accepted my answer and left. I went through this with him for 5 days. He finally got up and started walking around. I asked him where he was going. I told him I am going with you. I said you will not do any more drugs as long as I am here. He argued with me and asked what I was going to do about it. I told him "LOOK! YOU ASKED ME TO HELP SO YOU COULD GO HOME! YOU HAVE NOT EATEN IN 5 DAYS, YOU ARE LUCKY I MADE YOU DRINK WATER! I AM NOT LETTING YOU GO EVEN IF I HAVE TO STOP YOU PHSYICALLY!" Well 3 days before his flight he was pretty well cleaned up from the drugs and he thanked me. He then started to head to the flight line to catch a hop to Bien Hoa.

As part of the 1st Cavalry Standing Down, word has come to us that we would be going to other units. Even though my MOS (Military Occupational Specialty) is 11B, the head honchos were trying to put me with an aviation group. They mentioned to me, the 12th group, 17th group or 164th group. These are all air mobile squadrons. At the time I told my Mom and girlfriend not to write, I was in limbo of where I was going. During my time with A trp 1/9 my girlfriend was using my car to joy ride and pick up this guy she was dating. We had intentions of getting married when I came home from Vietnam. I got a letter from one of my friends that told me what she was doing. It was hard for me to believe so I wrote another friend and he confirmed what she was doing. I wrote my Mom and told her to sell my car. I wasn't sure I was going to make it home anyway. I loved that car. It was a 1965 Dodge Polara with a 413 hemi wedge engine. It was an overstock of cars used for police interceptors. The car was fast. My buddy Tom bought it right away. Even at Toms young age he knew how to make money. He sold the car to our friend Emmett for a profit.

CHAPTER 11

Tuy Hoa

I got my orders on June 10, 1971 to report to HHC 268th on a base called Tuy Hoa on June 12. When I arrived a day early I tried to find the 268th, nobody knew what I was talking about. This base is so big, It was like a city and who was a half a mile away did not know what other units were there. So I started to walk and walk asking soldier after soldier and still could not find the 268th. I finally saw an officer, someone with authority, I figured he should know. Well another dead end. He did point me to a series of hooches and said you can go there and maybe use their phone to make a call. That was easy for him to say. Who the heck was I supposed to call? Every hooch I went to I was told they never heard of the 268th. Finally after about an hour and a half, I saw a hooch with 238th PWC. I figured now this company is in the 200s they will surely know where the 268th is at. I walked into the Company HQ (Headquarters) and nobody is there. I waited for about a half hour and went into the hooch next to it and asked if someone was coming back to the HQ. I told him I was looking for the 268th. Nobody has heard of this unit. He said lets go back to the HQ hooch and said he had no idea where anyone was. I asked then where can I get something to eat. He pointed to 5 different places. Now my head is really spinning. I asked him which one would you go to. He pointed to the biggest building and said you can get hamburgers and pizza there and they are both good. He was helpful and I asked him where I could leave my duffle bags. I had two duffle bags, one was with my clothes and personal belongings and the other I had my M-60 broke down and a M-16 broke down. He said you could leave them in the COs office and I will let him know you will be back.

As I am walking over to this building to get something to eat, I noticed all the roads were paved with concrete curbs and sidewalks with street lights. I did not realize this building was about 2 blocks away. The closer I got I could see an outdoor theater and outdoor stage. I finally got to the building and it was huge. It has seating for at least 500 if not more. A live band was playing with female singers and dancers. The bar ran almost the length of the building which was about 75 feet from what I am guessing.

I went up to the bar and ordered a hamburger and a beer, I had to settle for a hamburger and coke because it was too early to sell beer. I began to think this place is going to be alright. I headed back to the HQ where I left my gear. I walked in and the Commanding Officer (Captains rank) and a Staff Sergeant were standing there. I asked where I could find the 268th. The Captain asked me for my full name and started sorting through a stack of papers and pulled the one out with my name on it. He said you are at the right place but we were not expecting you for 2 more days. This is when I found out that the 238th was attached to the 268th and now I am a 55B which is an Ammo Storage Specialist. So I asked the Captain is there somewhere I can stay until then. The Captain told the Staff Sergeant to show me to my bunk. Yes, a real bed. I have been on cots to this point and in a tent for 3 months. I got a foot locker and a tall locker cabinet. I put the duffle bag with the M-60 and the M-16 in my tall stand up cabinet. The next morning we were awakened for breakfast if you wanted it. I went the first morning to see where the mess hall was at. After that day I had nothing to do. The rest of the guys drank a lot. I did not drink because I wanted to make it home. The next few days I wandered the base taking pictures. Besides the Army being on this base, a division of the Air Force was also stationed here. I would watch the F-4s take off. There was even a C-5 Starlift. The C-5 was the largest military cargo transport plane. The guys also pointed out the beach area on the China Sea. I went there and saw guys swimming. I figured what the heck why not try the water. That night I cut a pair of fatigues to make swimming trunks. The next day there was nothing to do again, so I went to the beach. A group of 15 to 20 guys were already in the water.

I went in and it was warm and relatively calm. All of a sudden the 15 to 20 guys started yelling and running out of the water. They ran on to the sand and some started rolling around in agony. I thought to myself I better start getting out of the water. As I headed out, looking down, I saw 1 or 2 jellyfish. As I got into shallower water there were more right at the waters edge. If the waves would have been higher the jelly fish would have been washed on to the beach. I walked down towards the guys laying on the beach yelling in agony and there I saw the raised red welts on their legs from the tentacles of the jellyfish. That was the last time I went to the beach.

Picture was taken seconds before they were stung by jellyfish

There were only about 7 or 8 guys in this unit at the 238th. I am an Ammo Storage Specialist and have not seen any ammo or where it is stored. Very boring here. I kept looking for something to do. I have walked about the whole base. I tried to watch a movie but it was too crowded. In crowds with the units that I came from is where the enemy would throw a grenade or lob a mortar. It was the same in the club. I could not be around crowds. I am headed back to my hooch after lunch and trying to figure out what to do for the rest of the day. When I am about 5 hooches away from mine there was a huge explosion! I looked in the direction of the blast as I am looking for cover near the hooches, where there are barrels filled with sand for protection. The cloud of smoke and dirt that spewed into the air looked like a mortar round impacting. Then there was a second explosion. I am crouched down going from hooch to hooch when 2 guys walking on the sidewalk said, "You must be from the front." I said, "Yes, what is going on?" They said that they are blowing up old ammo. I did not say a word and headed right to the COs office. The CO and Staff Sergeant were in there talking and I said, "I need to get out of here and go back to the front where I belong." The Captain said fill out this paperwork but to let you know it will not do you any good. Then he said for me to leave his office and fill out the papers in my hooch. I started to fill out the paperwork and had it almost done. By this time I am bored and talking with the guys in my unit even though we had nothing in common. Their entire stay in Vietnam was having a sham job. No life or death situation or losing a comrade to hostile fire.

On June 22 I finished the transfer paperwork when the Staff Sergeant walked in and said you have to go to Bien Hoa. Here is your pass and the captain wants you out of here today. I said OK. Let me get my gear and the Staff Sergeant said you do not need anything because you are coming back here. The Staff Sergeant told me there are 2 flights to Bien Hoa and if I hurry I can catch one. I went to the flight line where there was only 1 Huey sitting there with its engine running. I went up to the pilot and asked where he was going, he said Bien Hoa. I asked if I could hop a ride with him. He said get on we have been waiting for you. He flew straight to Bien Hoa where a jeep was waiting for me. It was my friend from A trp 1/9 Randy Cramp. He became a driver for Colonel Ivey. I had written Randy and told him how unhappy I was with the 238th. Colonel Ivy had a lot of respect for Randy being an Aeroscout. They talked quite a bit and Randy mentioned I was not happy. The Colonel called me into his office right away. I let Colonel Ivy know how unhappy I was being with the 238th and about the explosions and need to go back to a Scout unit. We talked for awhile and Colonel Ivy said let me see what I can do. The Colonel told Randy if he needs him he would get a hold of him on the radio and for Randy to show me around. That night I bunked in the storage room across from the Colonels office. I was up early the next morning and the office was still locked. I went looking for a mess hall to get something to eat. Randy happened to be there and said he could not wait for me because he had to pick up General Putnum. I finished breakfast and walked back. I saw the office door open but did not go in. I went to my bunk and sat for awhile. Randy came in and said the Colonel told him to look for me because they were waiting for me. I asked Randy who is waiting for me. Randy said General Putnum and Colonel Ivey. I said oh boy this has to be good. I knocked and was asked to come in and was told to take a seat. General Putnum said I hear you are unhappy with the 238th and can you tell me why? I told General Putnum about the paved roads, sidewalks and the demolition of old ammo. The General laughed and said I have got it now. He then said most soldiers try to get these jobs to have it easy in Vietnam. Let me see what I can do. Later that day Colonel Ivey called me into his office and told me he found a unit for me. He said A trp 3/17 1st Aviation Brigade needs experienced door gunners for their Scout Platoon. I told him I will take it, I just need to get out of Tuy Hoa. Colonel Ivey said I will call the CO at the 238th and tell him you have been transferred. When you get back to Tuy Hoa, pack your belongings and find a flight to Lai Khe and report to the Commanding Officer and your paperwork will catch up. Well I took my time getting back. Spent another day with Randy. We went to the Mexican Restaurant La Casa Grande. Another good meal. This time they did not have the hot cherry peppers or any hot peppers.

Before heading back to Tuy Hoa I went to the flightline to try and catch a hop to a Marine Base in the Quang Tri Province called the Rockpile. My cousin Joe Konicek was stationed there with an ARVN (Army of the Republic of Vietnam) Advisory Unit. It took three helicopters flights to get me there but I made it. The chopper pilot called ahead to see where the best place to drop me off to find Joe. Luckily we were instructed to drop me off on the top of the Rockpile where Joe's last location was known. Joe had already been in Vietnam for two years and seemed to have no intention of going home. As the helicopter is dropping me off I could see my cousin with an M-79 grenade launcher in his hand firing at a jeep below the hill. I said hi and he said in a very calm voice hi back. Then I asked him what are you shooting at?, he said I am trying to kill my Colonel for almost getting me killed on a mission. We talked for a while and he said you might want to try and leave today because intelligence said they were going to be hit (attacked) that night. Before dusk I figured I better start looking for a way out of here, lucky as usual I found a flight out to Qui Nhon. After that incident with the Colonel, the upper command of the Marines figured Joe should be sent back to the United States to Camp Pendleton, California. Shortly after getting to Camp Pendleton Joe was released from active duty with an Honorable Discharge.

My Marine cousin Joe at the Rock Pile

I got back to Tuy Hoa around June 28th only to find out the guy in the bunk ahead of mine died. He died of an alcohol overdose. A ten dollar bet was made that he could not chug a fifth of vodka. He won the bet but passed out and drown in his own vomit. That was more conformation I wanted out of this company. In a combat unit you do not have time for stupid shit. As I am packing my gear I saw my second duffle bag was gone and I asked where it was. In the service you do not mess with other peoples shit. It is a good way to get hurt. They said they knew nothing about it, I responded by saying I suppose it just disappeared. As I continued to pack, the CO and Staff Sergeant walked in and the CO started yelling at me. The CO yelled, "YOU WENT OVER MY HEAD!" I turned and said firmly, "I TOLD YOU I WANTED OUT OF HERE!" I turned back around to finish packing. The CO yelled again, "DO NOT TURN YOUR BACK ON ME!" I had finished packing, turned 90 degrees and threw the shoulder strap from my duffel bag over my shoulder. As the duffel bag swung across my back the Staff Sergeant stopped it from swinging into the Captain. I asked about my second duffle bag, the Sergeant said, "We confiscated that and we could bring you up on charges." I said nothing, turned and headed for the door. The Captain yelled, "YOU GET BACK HERE!" I kept walking and went through the screen door. I heard the screen door slam shut and could hear the Captain yelling at the Staff Sergeant, "STOP HIM!" I started to walk down the sidewalk to head to the flight line. I could hear the screen door slam shut from the hooch I came from. I glanced back and I could see the Staff Sergeant standing on the side walk and the Captain barking at him, "AREN'T YOU GOING TO STOP HIM!?" The Staff Sergeant finally said to the Captain, "WHAT DO YOU WANT ME TO DO?, TACKLE HIM?, HE IS LEAVING!" I totally ignored them and kept walking. I was able to catch a flight right away.

Thanks to General Putnum and Colonel Ivey I arrived at Lai Khe on June 30, 1971 to join up with A trp 3/17 Scout Platoon. The first person I saw after getting off the Huey as I walked from the flight line and into the Scout area was Captain Knight. I flew with him with A trp 1/9 before they stood down and took their colors home. Captain Knight was walking with other officers and in a loud voice said, "Hey! There's Gregory!" He came up to me right away and asked what was I going to be doing here. I told him as far as I know I am going to be a Scout. Captain Knight asked me what my MOS is and I told him 11B. He then said we will get that changed and you will be flying Scouts. Captain Knight knew I would be a good Scout working with me at A trp 1/9. He knew I was a good observer and gunner. My MOS was changed that day to 11D2F. 11D2F is still infantry. The D is Recon and the F is Flight Clearance Status.

From the first day at Lai Khe I was treated well. The very next morning I was flying with Captain Knight. We made a good team at A trp 1/9 and now at A trp 3/17. I really was not prepared to fly the very first day. I had to use someone else's equipment. When we got back I was told to keep the M-60 machine gun and chicken plate because the previous gunner quit. This was just a starter kit for me. I had to build my own weapon system and arsenal. I had more freedom with what I carried in the OH-6. Every night after our missions I would gather ammo boxes, secure fragmentation, white phosphorus and concussion grenades. I had another ammo box for my M-79 HE rounds, CS gas, incendiary and smoke grenades and of course 1 ammo box for approximately 7,500 rounds of 7.62 M-60 machine gun ammo. We were also semi-allowed to build our own bombs with C-4. It seemed like everybody's family sent goodies in coffee cans. In a 1 pound coffee can you could get around 3 sticks of C-4. That is about 3 and 3/4 pounds of plastic explosive. Some guys put nuts and bolts mixed in with the C-4. For the fuse we used the fuse from a fragmentation or concussion grenade. When you dropped them into a bunker it was guaranteed to lift the lid and expose the logs supporting the dirt on top. It gives you a good idea how big the bunker is.

I began to fly with Captain Knight on a regular basis and being that he was the Scout Commanding Officer, it was his decision on the missions he flew and the hours he wanted to accumulate. With Captain Knight I would get First Light around our compound, 3 missions in our AO (Area of Operation) and Last Light around our compound. We flew together most of the time. His flying was a little erratic and speed was a little fast on the continuous right hand turns when reconning an area. Then when something catches his eye to the left he spins the nose of the aircraft to the left. Now I can not protect the aircraft or crew. Making a left turn puts me in an upward position. When I am in this position I can not get my M-60 machine gun to fire down when I am looking at the sky. Then back and forth erratic movements, taking me out of position again and again. One time we were taking fire and I had to let go of my M-60 machine gun to hang on. The speed and zig zag pattern to evade enemy fire kept me from doing my job. It is a good thing my M-60 machine gun was attached to twisted tie wire with D-rings, otherwise it would have gone out the door.

Helicopter Mr Schwab was shot down in by small arms fire.
No injuries to all 3 aboard

Working on my aircraft 238

From the left, SGT Brown, WO1 Schwab, SP4 Gregory and PFC Jeffers

Finally I got a break. I started to fly with WO Richard Schwab and Captain Skinner. As far as I was concerned these were the best pilots. WO Schwab was a little better than Captain Skinner but their styles were similar. WO Schwab would at least talk to you about what was going on the ground and ask for your opinion. Mr Schwab, as WO (Warrant Officers) are called, was a very level headed, intelligent pilot. When he sat in the Kevlar pilot seat he became one with the aircraft. He flew by the seat of his pants and always kept my machine gun in a position to protect the aircraft and crew. The only drawback of the 2 pilots is when I got into a gun fight they wanted to watch. They actually brought the aircraft closer to the fight. I am shooting left, right and center as I see the enemy either running or turning towards me to try and get a shot off at us. In the back seat you can not fire and push the button to talk at the same time. One hand is on the trigger and the other hand is pushing down on the M-60s upper heat shield to steady the gun as you are shooting. There were times when I was firing and sweating profusely in a gun fight and wished I could press my mic button to say get the hell out of here or punch them in the back of the helmet.

CHAPTER 12

68-17257 Down

On July 24, 1971 I was not scheduled to fly. I had the maximum number of hours allowed to fly combat missions. Some time near or around 1630 hours we got a call to scramble because one of our aircraft number 68-17257 was shot down. Some of the aircraft were low on fuel and could not stick around the crash site. I grabbed my M-60 and helmet but forgot to take my chicken plate (chest armor). Mr Schwab happened to be down also and was my pilot for the scrambled mission. We took off and I did not want to tell him I left my chicken plate. It was about a 40 minute flight to the crash site inside the Cambodian border. When we got to the crash site there was a Huey hovering and moving towards the downed aircraft. I could see someone was being lowered through the triple-canopy jungle. It looked like the medic was riding on a jungle penetrator, but I found out later it was a log tied to a rope. Mr Schwab and I reconned the area to keep the rescue medic safe as he was all alone. The foliage was so thick that even our rotor wash could not separate the vegetation to see through. I could not fire my weapon because I did not know if our guys were running or even where the aircraft was at. The Huey that came for the rescue was not from our company. He happened to hear the distress call and came to offer his assistance and happened to have a medic on board. The wind was gusting making it difficult for the Huey to hover.

Mr Schwab and I are doing what we can to keep the Huey and the medic safe. The Cobra circling over head spotted approximating 200 to 300 NVA headed towards the direction of the crash site. As the medic was carrying the pilot WO Timothy Wiltrout with a gun shot to his leg was taking sporadic gun fire from NVA ground forces near the crash site. The medic held on to WO Wiltrout and was carried off on their makeshift penetrator to a safe site to get both on board the helicopter. There was no winch on this helicopter so they had to be lifted out and set down to climb aboard. He patched Mr Wiltrout and then headed back for Sergeant Gregory Antunano observer and SP4 Randall Dalton doorgunner. From the medics report they were already dead. In an effort to recover the bodies the ground to air fire became intense. The NVA had moved to the area in force. There was a lot that hampered the rescue mission. The wind was gusting, the sun was starting to set and the fuel became an issue for the Huey and the Cobra. The call came in to abort the mission. The Huey had no choice to pull off because of heavy ground to air fire and lack of fuel. Mr Schwab and I kept looking for movement and I guess a false sense the crew would suddenly emerge into an opening where we could pick them up. We kept probing the area even to a point of being dangerously slow. The Cobra stayed with us even though he kept telling us the mission had been aborted and to pull off. Mr Schwab never once responded to the Cobra pilots request. The Cobra was running low on fuel and called to us that he was returning to base, again. Mr Schwab gave no response and I sat in the back silent. I watched the Cobra make a sharp turn and head for home. Mr Schwab gave me an order to keep my eyes open. It was part of my job anyway, but I knew he was looking for something that is not there. We were all alone with no cover fire so my alertness peaked to keep us safe. After about 20 minutes and just enough fuel to get back Mr Schwab asked if I was ready to head back. My response to him was lets go home. Again the silence. There was no radio chatter or talking between Mr Schwab and I. The silence and knowing two of our guys were still on the ground made the ride back the worst and loneliest ride ever.

When we got back we said nothing to each other. Mr Schwab shut the chopper off and walked away. I stayed back to make sure the chopper was secure and my equipment was put away. I walked to my hooch and was told we were going to have a debriefing at 2000 hours. I took a shower then I was bombarded with questions by our Blues Infantry Rifle Platoon and by my Scouts asking me why we left Antunano and Dalton. I told them Mr Schwab and I only caught a glimpse of the crash one time.

The triple canopy made it difficult to see anything. When I told the guys the mission was aborted they were angry and walked away. My anger was building like what could I have done. Then the grunts came back and said what do you think about volunteering to go with us to get them? They knew that I was an 11B and well trained on the M-60 machine gun. They said they had a Huey pilot that was willing to take us. I said right away lets do this. Prior to the debriefing I heard a lot of commotion and swearing outside my hooch and a Sergeant throwing things. I said, "HEY KNOCK IT OFF, STAND DOWN!" He looked at me, grabbed my shirt and started crying. He said the Motherfuckers denied us to go get our men. He then said I feel like killing someone. Luckily I was able to settle him down and talk sense to him. I told him come to the debriefing because I am going to have my say and I want you to hear it. He said, "OK Vern, I will for you." Shortly after we went to the debriefing held in the mess hall. The Company Commander Captain Robert Rice gave a speech telling us what happened and why the mission was aborted. The more he spoke the angrier I got. At the end he asked if anyone had a question. So here I go when I asked, "why weren't our blues sent in?" His response referred to the time of day. He said by the time they got there it would have been dark. I told him I scrambled with a pilot and spent over 40 minutes on station and by the time we got back it was still light. He said it was a decision he had to make because of the distance the ground troop would have had to make to even get to the crash site and the possibility of losing more men. I then shouted, "Why was our request to volunteer for the mission denied?" We had at least 8 well trained grunts and a pilot to take us. The only answer I got was that it was his decision to not lose any more men. At that I turned and walked out.

The next morning our Blues saddled up and boarded the Hueys. The Scout unit went out with 2 OH-6 Loaches, 2 Cobras and our Huey C & C Chasebird. As one Loach recons the crash site the other Loach stayed further out looking for enemy still in the area as the Cobras stayed perched high over head ready to pounce at the first sign of trouble. Once we declared the area secure, the Rifle Platoon moved in. It did take some time for the Rifle Platoon to get to the crash site just as the Captain stated at the debriefing. They found the aircraft had been stripped and the bodies of Antunano and Dalton were nowhere to be found. Everything came into question now. Was the medics findings correct? He was not from our unit so we had no way to ask him questions. With the bodies gone were they alive or dead? A lot runs through your head. A thorough search of the area was done by the Lowbirds, the OH-6 Loaches, to either find our guys or someone to take our revenge on. It was strictly jungle so there were no civilians in the area. The Rifle Platoon was then extracted safely.

The next day our missions continued. No time to grieve or feel bad. Numbing sets in and death becomes a way of life. Our missions were in the Iron Triangle and Cambodia. Mostly working the rivers, channels and dense jungles of eastern and northern Cambodia. In Vietnam we also worked trails that came off the Ho Chi Minh Trail. The Ho Chi Minh Trail was a dirt road/path from North Vietnam to South Vietnam crossing through Laos and Cambodia. This was their super highway for supplies. We did everything we could to disrupt their supply chain.

Just as we felt things are settling down with the loss of our crew and we are back with operations, we get a surprise Sapper attack on July 29, 1971 at our home base Lai Kai. The Sappers snuck in somewhere between 0230 and 0330 hours. They blew up 4 aircraft and damaged a fifth. We did not know if they set satchel charges or put rubber bands around hand grenades and dropped them into the helicopter fuel cell. The explosions woke everybody. We all started to run towards the flight line with weapons in hand as we got closer rockets and ammo started to cook off from the burning Cobras. It was quite the light spectacle with rockets igniting, taking off and exploding. 20mm rounds cooking off and tracers piercing the darkness. As daybreak came you could see the devastation. Two OH-6 Loaches and two AH-1 Cobras destroyed and one Cobra damaged. My helicopter number 238 was untouched. The Loaches had no ammo on board. We would take it off each night. This disrupted our operations for the next week. The pieces of helicopters had to be picked up. Every inch of the flight line had to be walked and cleaned so nothing could get airborne by rotor wash and damage other helicopters or hurt someone by flying debris.

Remains of 3 of our Loaches. Cobras were also lost in the sapper attack

Helicopters started to come in from units standing down. Pilots were taken from Lai Khe to pick up these choppers at Bien Hoa Airbase. A quick check of the new choppers were done and put into service right away. Our missions resumed around our base first. The Sappers were sent into our compound by an NVA regiment close by, according to intelligence reports. We never did find them but we had plenty of tracks and trails to follow. They stayed concealed well.

CHAPTER 13

Rice Cache

On August 31, 1971 PFC Frederic C. Saunders, Jr. died on a mission to destroy a stockpile of rice we found and secure any weapons that may be found also. The day before we were doing an aerial reconnaissance of an area suspected of enemy movement and concentration. As we flew at treetop level, I saw what I thought was black plastic sticking through the dirt. I shot it up with my M-60 machine gun and uncovered what looked like rice. The pilot then said to drop a fragmentation or concussion grenade on the suspected area. I dropped 3 grenades and uncovered a stockpile of rice. I then marked the area with smoke so the Cobra gunship could roll in and hit the area with rockets. The Cobra gunship made 2 passes, knocking down the thick foliage and uncovering more of the rice stockpile. The cache of rice was larger than we thought. I threw more fragmentation and incendiary grenades and did very little damage. We then got a call that our Blues Rifle Platoon will be deployed tomorrow to destroy the cache of rice. On the eve of the 31st PFC Saunders came over to our Scout hooch. We were just setting up, getting ready to play cards. Saunders said I would like to say goodbye to you guys. We did not know what he was talking about at first. I stood up and he shook my hand and said you are a good guy and a good friend. He then shocked me by saying, "we are going out tomorrow and I do not think I am coming back." That hit me hard. I said, "no way! You are coming back!" At that Saunders got a little agitated, turned and abruptly walked away. I went straight to the Officers Quarters and Saunders was there saying his goodbyes. I demanded to work that area to make sure it would be safe for our Blues Platoon. We all let it go at that point and went on with our regular routine.

The morning of the 31st I got first mission into the area of the rice cache. Nothing had changed. We worked the area around the cache and there was nothing to be found. After about 1 1/2 hours our relief came in and reconned more of the area and taking a second look at what we had already gone over. The all clear was given to A trp 3/17 Blues Platoon to go ahead with the insertion. No more thought was given to what Saunders had said. The Blues Rifle Platoon had a 2 mile walk in the jungle and along old bomb craters. The bomb craters were huge. Some of the craters were 30 feet across and 25 feet deep filled with water. The path they chose or were guided by aerial support, either way there was no good route to the cache site. The path they took was only about 2 feet wide and went right along side one of the larger craters filled with water and a 2 to 3 foot berm of dirt to the other side of this trail. I was on the next Scout Team to go up to give cover to the Blues and relieve the crew already on station. Then the call came in. One of the Blues slipped into a crater. With all the ammo a grunt carries it was sure to take him to the bottom. PFC Saunders carried at least 21 magazines of M-16 ammo, fragmentation grenades and 2-100 round belts of M-60 machine gun ammo crossed over his shoulders and across his chest. I came on station to see guys in his platoon diving with only their underwear on trying to find him. Not a single guy diving could reach PFC Saunders. We were then told the Blues Mission is being aborted. Then we were ordered to go to the cache of rice and throw all of our incendiaries onto the stock pile and for all replacement teams to do the same. We were told to prepare for a recovery mission on September 1st. The next morning the officers could see I was visibly upset. I felt as if I did not keep PFC Saunders safe. I was not allowed to go on the first or second mission. I was able to go on the third mission. It was not until part way through the second mission the diver found his body. They brought his body up to the surface and placed it on the edge of the crater. You could see his platoon comrades crying and some of them vomiting. We could not get too close because of our rotor wash. We needed to stay focused and recon their perimeter to assure no enemy could attack while they were recovering the body of PFC Saunders. That night there was a lot of drinking and cussing with the guys in the Blues Platoon. They lost a good guy.

CHAPTER 14

Joint Operations

Missions began to intensify reconning areas with more and more enemy activity as FSB (Fire Support Bases) began to close. This gave the NVA more lateral movement without being attacked or shelled from artillery batteries. We started working the Central Highlands. That was hairy! Up and down steep slopes. Watching out for flocks of parakeets. Monkeys with human like features that were not afraid of us. They would climb to the top of the trees and reach out towards us. This is where I saw a booby trap for helicopters. It consisted of 2 long ropes crisscrossed and draped over trees leading to 55 gallon drums at the ends of the ropes and one 55 gallon drum in the middle. The 55 gallon drums on the corners were tilted inward slightly. They had an explosive charge underneath and the barrels were partially filled with rocks. We got lucky that it did not go off but we did get a good look at the booby trap. Our rotor wash was to move the tree branches and the ropes were to trigger the explosives to knock us out of the air. The Central Highlands were tough to recon. You had the steep slopes on both sides going down to a narrow valley below. The slopes had thick vegetation and we could not use our rotor wash to blow the foliage open to see what was underneath. Our Platoon Commander made a decision to stop reconning the Central Highlands. It was way too dangerous. We were put into a situation where the enemy had the advantage.

September 1971 our missions began to change. We started to recon bunker complexes prior to B-52 strikes. Then on the same day we were told to back off 5 miles in a specific direction. Then, within ten minutes the B-52s start their bombing run. Once the B-52 pilot tells us the last bomb has impacted, we are allowed to go into the area fifteen minutes after. We entered the area as the dust is settling and assess the damage done to the bunkers. Huge craters are created from the bombs. Bunkers are totally eliminated while others are damaged beyond their usefulness. We also got to watch the aerial ballet of the Douglas A-1 Skyraiders. We would first find a smaller bunker complex and call in for aerial bombing. On one mission in particular that I was involved in, two Skyraiders were called in to give us support. We would first mark the area with smoke so the Skyraiders would know where to bomb. The first Skyraider goes into a dive to the target area, firing rockets and as the plane starts to pull out of its dive, it drops bombs. The second Skyraider is on its dive as the first plane is pulling out of its dive, firing rockets to give the first plane cover. When the second plane is pulling out of its dive, it is dropping bombs and as he pulls up, the first plane is already inbound, firing rockets to give cover to the second plane. It is a well choreographed flying effort, not only to get the job done but to protect with cover fire for a successful mission. This is done three times, then we go back to assess the damage. If more needs to be done, they have enough bombs and rockets for 2 more runs. We did this for approximately one week. Then moved on to recon brush areas and canals off of main rivers.

Headed in to assess bomb damage

Clumps of brush or bamboo is where bunkers could be found. On September 15, 1971 we were finding nothing, no foot prints, no food wrappers. Towards the end of our mission we get a call that Two Six was inbound to relieve us. Then we spotted bunkers and I started to fire my M-60 machine gun into the entrances. It did not seem to have any activity here so we marked it for the next crew. We started to climb to altitude and did not see our relief team. As we are headed back to our staging area, we passed the relief team which was over 5 minutes late. By the time they got to the AO (Area of Operation) the smoke was gone. We guided them to the bunker site by using land marks we remembered. That is how you get back to where you saw something. We are about half way back and get a call that Two Six crashed. We had plenty of fuel and headed back. By the time we got there the Chasebird (C & C) picked them up. We had no idea what caused the crash as we reconned the area for bad guys. Then we get a call to check the aircraft if anything is still on. My pilot set down and let me off. All switches were still on so I went though shut down procedure. Aircraft #68-17175 was laying on its side. I had to hop up to climb inside to shut everything off. We then left the area and was ordered not to use white phosphorus grenades until further notice. As Two Six got into the area, they figured they would mark the bunkers with white phosphorus so the Cobra could roll in and fire rockets to destroy the bunkers. The crew consisted of Pilot Captain Herbert Skinner, Observer WO1 Edwards and rear gunner SP4 Charles Jaquins. When SP4 Jaquins went to throw the white phosphorus grenade it had a short fuse and went off instantaneously when the spoon came off. Captain Skinner had white phosphorus burns on his neck. SP4 Jaquins suffered the brunt of the explosion. He was rushed to 24th evacuation hospital in Long Binh. He was immediately transported to the Burn Center at Fort Hood, Texas. It was not more than a week later we learned he had succumbed to the burns and died.

Our Scout missions became more diverse towards the end of September. We started working with the Brown Water Navy in the Delta. The only contact we had with the River Rats was through our C & C (Command and Control) Huey flying at 1,500 to 2,500 feet with the Cobra circling the Scout helicopters position. We would start out by flying over their moored boats to tell them we are on our way to clear the river banks of ambushes. We could only recon one river bank at a time. As we, in the OH-6 Lowbird, would recon the most obvious to us through experience areas for enemy presence. On any mission we have three pairs of eyes. The pilot sits in the right front seat flying the aircraft in slow clockwise turns moving ever so slightly forward. The gunner, equipped with a M-60 machine gun, an assortment of explosives and incendiary devices, sits behind the pilot in the rear compartment, on a chicken plate laid on the floor with his legs hanging outside. The observer sits in the left seat next to the pilot watching the left side with a M-16 and a smoke grenade ready to be dropped on command or when we are being shot at. This particular day I saw a lot of footprints heading towards the river bank where there were several bunkers. I started to shoot my M-60 machine gun to try and draw fire. They know if they panic and fire back they could have hell rained down on them, by either the Cobra rolling in with rockets, 7.62 minigun or 40mm grenades from the automatic grenade launcher on the Cobra. If that is not enough, artillery or air strikes can be called in. No return fire was taken so we came in slow and I threw fragmentation grenades into the bunkers. Still no return fire from anywhere. We had the Cobra roll in and try to uncover a bunker or two with its 2.75 inch rockets. As the Cobra rolled in the River Rats were coming down the canal at full speed. We were told to get altitude and out of the way. The River Boats had more fire power and wanted in on the action. As soon as the river boat got into position they started to take gunfire. A gun fight ensued as our relief team came on station. As we left the area we could hear the radio chatter and the Cobra rolling in on the enemy positions and the enemy running. This area did not have dense foliage. It was mostly farmland with small clumps of trees and large pineapple type plants along the river banks. Then as we were still listening, on our way to fuel up, our relief team was told to pull off. An infantry unit was on its way by boat to clean up. After refueling we were sent back to make sure the infantry platoon had aerial cover to make a successful extraction. We only worked with the Brown Water Navy for a couple of days. Then we went back to the jungle recon.

It was only a few days later, when we were called to do another joint operation with the Navy SeaWolves. SeaWolves were an elite unit of the

Navy usually stationed by river bridges. They also had "hand me down" Army B-Model Huey gunships. We kept going there and parking our OH-6 and it seemed like nothing was happening. I got to know the guys with the SeaWolves. So then the trading began. They wanted a Colt 45 which was my sidearm at that time and I wanted a 38 revolver. I gave him my Colt 1911 semi-auto 45 and holster and he gave me his Smith and Wesson revolver with a real cool western style holster. I could get 38 caliber rounds but he couldn't get 45 caliber rounds. I told him we'll be back tomorrow and I will bring cases of 45 ammo for him. Then as we talked standing by my aircraft they saw I carried white phosphorus and fragmentation grenades. They asked me if we had any extra white phosphorous and fragmentation grenades because they only had concussion grenades. They used concussion grenades to throw into the water at the base of the bridge every once in awhile in case a Sapper was nearing the bridge under water. I said when we are ready to leave, you can take what I have and I will bring back cases for you tomorrow. The next day came and I loaded two cases of fragmentation grenades and two cases of white phosphorus grenades and one case of 45 ammo. My unit ran out of concussion grenades and it was my understanding they were not going to order any more. The SeaWolves asked me what I needed, I told them we needed concussion grenades. On this particular day we already knew we were not going out on the mission again. I had to wait for the pilot anyway so I told them I'm going to the mess hall to get something to eat. When I got back the pilot was already sitting in the aircraft and the SeaWolves had loaded the back of the OH-6 with case after case of concussion grenades. It was so loaded I had nowhere to sit in the back. We got the order earlier to return to base and I told the pilot we might be overweight. I had to take out three cases of the concussion grenades so I'd have a place to sit. The pilot then told me, "let me try to get off the ground." Well we were red lining and barely got one skid off the ground. I had no choice but to off load some of the cases of concussion grenades. The day before we were able to take three cases of concussion grenades. I now have to take off seven cases of concession grenades to get down to the three cases. A couple of the Navy SeaWolves asked what was wrong. I told them we were overweight and could not get off the ground. He asked if we could hold on a minute because he has something for us. So

I did everything I could to stall the pilot from taking off. Then I saw him coming and the pilot was getting aggravated with me. I said, "hold on - a guy is walking up with something." The SeaWolf was bringing something and the pilot said to me, "Hurry up!" He handed me a box loaded with whiskey. I placed it in the back and didn't say anything to the pilot. That night I split the whiskey up with our "Blues Platoon Grunts". These are the guys that come to get us when we are shot down.

Day three at the SeaWolves base and it looks like the mission is a go. My team on the OH-6 had to wait for the SeaWolves B-Model Huey gunships. We sat the chopper down but never shut down. Then the gunships flew directly over the top of us. There was no mistaking the sound. We took off following their orders. Once we got to the AO (Area of Operation) we were guided into the bunker complex by the gunship commander. This area had brush and very few trees so we were able to get about 20 feet off the ground. I was able to see movement in the bunker entrances and started to fire my M-60 machine gun into the entrances. Evidently these gunship pilots thought that we were in trouble and started inbound with their gun run. They yelled over the radio that we were to move out of the area immediately as they were inbound. We didn't think that was an emergency to move too quickly. That was a mistake. Just as we are moving out of the area I could see the front of the gunships at low level firing rockets and the door gunners firing their M-60 machine guns. Too late for us to get out of the way, then we heard them say they were inbound for a second pass. Then my pilot said we better get clear of this area. As the gun ship passed over the mercenaries on their second gun run the mercenaries began to run towards the bunkers in pairs. The dust did not settle from the rockets impacting as the mercenaries ran into the bunker complex. The enemy was doing everything they could to escape but from what I saw there were no enemy survivors. Enemy count had to be around 50 with my best guess. No mercenaries were killed. I did see one mercenary with a minor flesh wound. Then we got a radio call from the SeaWolf Commander that our day is done and thanked us for being there and said that we could return to our home base.

CHAPTER 15

Working With Allied Forces

In October 1971 we went back to jungle reconnaissance and reconnaissance of LZs (Landing Zones) prior to allied troop insertion.The United States Army was pulling back and turning more missions over to The Army of the Republic ofVietnam.

On Otober 4 we were to recon an LZ. As we checked the tree line I could see enemy movement in bunkers.There was no time to tell the pilot I was opening up with my M-60 machine gun as the NVA tried to fire their AK-47s at us. Fortunately the way they designed those bunkers, they could not fire their automatic weapons up at us. They started to bailout of the bunkers and run as I fired my M-60 machine gun. We followed at the top of the trees and lost them in the thick jungle foliage. We had the allied force do a go-around until we were sure we neutralized the enemy presence. It did not take too long and we gave them clearance to insert. We stuck around to make sure the Hueys got in and out safely. Then we moved into the jungle to see if the enemy was setting up an ambush. We stayed as long as possible until we ran low on fuel. No relief was coming in for this mission. Our job was done and we returned to our staging area where the rest of the Scout unit waited. I received an Air Medal with "V" Device for this combat mission.

On October 14 our day was ending. Not much was going on in our area of operation that we were working. Then as we were headed back to our base camp an emergency call came in from The Army of the Republic of Vietnam Unit 4-8 Battalion in contact with a North Vietnamese Army Battalion. The fighting was fierce, intense and close air support was necessary. The allied force ran out of smoke grenades to mark their location for close air support. We were asked by our Commanding Officer Captain Rice if we had enough smoke and fragmentation grenades to go on a dangerous resupply mission. We did and then we were asked to go voluntarily. All on board had to agree or the mission would have to be aborted. Captain Skinner was the pilot, PFC Phinney was left seat observer and I was in the rear compartment door gunner behind the pilot. The ground to air fire was intense. We hovered above the lead element and I dropped all my smoke, fragmentation, concussion and white phosphorus grenades as fast as I could. As we were unloading, the Cobra was making gun run after gun run, firing their minigun and rockets at the hostile force in the tree line. Once we offloaded Captain Skinner turned our aircraft to his and my side towards the enemy positions. I began to open up with my M-60 machine gun as Captain Skinner moved towards the enemy force. They began to retreat as I fired and the Cobra attacked. Then fuel became an issue and we left the area secure for the time being as Blue Max, an attack force consisting of a 2 Cobra Attack Team came back on station in support of the 4-8 Battalion of the Vietnamese Army. For these actions I was awarded my second Air Medal with "V" Device. From this mission I also received an individual award from the Army of the Republic of Vietnam which was The Gallantry Cross with Silver Star.

Sometime in August I started to get letters from my girlfriend infrequently. I kind of knew something was wrong. I thought she got hurt or died. So I asked my Mom to check up on her and ask if she is ok and to please write. All of a sudden the letters started to come from her again and I thought everything was ok. Then she writes and said that she really did not have much to say and hopefully she will get a letter out once a week or one every 2 weeks. That was fine as long as I knew she was ok. Nothing really changed between me and Candy I thought. Then the letters stopped again. On September 29th I hadn't received a letter from Candy for three weeks. On the same day my Mom said in a letter, Candy brought my ring back. At this point I knew it was over but no official Dear John letter. Then on October 5th I got my Dear John letter from Candy. She said that we should start to see other people. (Apparently she was unaware that I knew she was picking up her new boyfriend in my 65 Dodge Polara) Evidently she did not know there are no women on the front line. The largest base I was on after Phuoc Vinh was 165 combat men. Sure, we had cooks, clerical and supply but if the shit hit the fan they were supplied with M-16 rifles. There was not a single woman from the United States on these bases. We did have hooch maids that I would see once a month on my day off. I would leave at day break and not be back until sunset. The hooch maids could not be there after dark. I wrote Candy and wished her the best and hoped I would take a lucky bullet. Other guys that got Dear Johns committed suicide. They would shoot themselves or take a helicopter on a one way ride. One guy in particular while I was there took a CH-47 Chinook up and started to do barrel rolls. This helicopter is not designed for that. It was amazing, he deliberately did one barrel roll and as he went into the second barrel roll, it plummeted to earth, crashed and burned. I still had plans for my future and would not intentionally hurt myself.

At the end of October and into November we were assessing airstrikes. We would enter the area 15 minutes after the last bomb had exploded. The B-52 airstrikes were the best to watch and the size craters they left. The one thing I have to say is that they are not very accurate. We at A trp 3/17 found a large bunker complex. It was approximately two blocks long and several hundred feet wide. The B-52 airstrike caught the corner of one end. It did a lot of damage and uncovered some of the bunkers. If the bomb did not kill, the concussion had to have taken many lives. The ground is covered in a thick dust from the bombs. The bombs made craters, from our estimate, 25 feet across by 20 feet deep at the deepest point. The craters had the resemblance of a cone or funnel.

CHAPTER 16

Delay And Destroy

It was announced in November that our missions were to "delay and destroy". As the United State Military continued the pullout of our troops, the job was to disrupt any movement and keep them from amassing supplies and combat troops. Our AO (Area of Operation) was closer to our home base, Lai Khe. There were times we were sent out further if intelligence tells us where enemy supplies might be or enemy are concentrating. Our base came under yellow and red alerts on a constant basis. Intelligence was following enemy troop movements and we happened to be right in their path. Yellow alert meant potential to be attacked was high in the 75% range. Three nights in a row we had a red alert. Intelligence had a report that a brigade (approximately 3,000) of NVA combat troops was headed directly towards our compound. Extra Blues, our infantry riflemen were put on a perimeter guard duty. At dusk artillery fired marker rounds at our perimeter in case they had to fire for effect with little notice. The 105mm Battery that was to give us cover fire had an assortment of rounds they could use. They had HE (High Explosives), white phosphorus and Beehive (Flechette) rounds. The flechettes were #1 common nails with it having 4 fins rather than a flat head. The 105mm round had approximately 8,000 of these darts.

After the third night of red alert we lucked out. The NVA went around us to the west. One of those nights we had a new guy added to our scout unit. His name was Ken and he got the back room of our hooch. My room was the first room to the right as you walk in. This is where we played marathon spades most of the night. Last night the siren went off and it was the "gooks in the wire alarm". We saw him run past with his

M-16 and his bandoliers of ammo, dressed in OD (Olive Drab) green socks with combat boots, OD boxer shorts and an OD t-shirt. We heard the screen door open and slam shut and a few seconds later Ken opens the screen door and stands in my doorway. He said to the four of us as we continued to play cards, "didn't you hear the alarm that there are gooks in the wire?" We said, "yes let us know when the war is over." He then said, "You guys are crazy!" He was about to find out how crazy Scouts are because his first mission is in the morning. He came walking back about a half hour later with mud on his boots and clothes. We knew nothing was going on. No rifle fire or artillery rounds impacting. We have been in country in combat for over 10 months so we knew what to listen for. We made him clean his rifle that night because he got mud on it. He said he would do it in the morning but we demanded he does it now. We explained that not only your life depends on that rifle but ours also.

CHAPTER 17

R & R

R & R (Rest and Relaxation) was coming up for Donald and I. Don had the same job as I did, a back compartment door gunner. We put in for R & R a day apart and by going to different officers so they could not stop us from going together. We were hiding out November 15th and as we were boarding the mail chopper an officer asked where we were going. We told him Bangkok. He said we can not let experienced gunners go at the same time. We totally ignored him and climbed on the aircraft. We were on our way. We took another jump from Phu Loi to Tan Son Nhut to catch a commercial airliner to Bangkok, Thailand. At the airport we met up with a grunt from the 9th Infantry Division. Once we got to Bangkok we took a cab from the airport to the hotel. A bellhop showed us to our rooms. He asked very politely if we needed anything. Don right away said, "Booze!" I said, "Some beer." Bob from the 9th infantry asked for some marijuana. The bellhop nodded his head and left. We got settled in our rooms and agreed to meet in my room before we headed out. There was a knock at the door. It was the bellhop with Budweiser, a bottle of Jack Daniels and a pack of marijuana that was quite interesting. The Marijuana came in a 20 pack just like cigarettes. The box, if you looked at it quickly looked like a Marlboro pack. The red and white color on the pack was identical. The only difference was the height of the pack which was shorter. The Marijuana cigarettes had a shorter filter. It was about half the size of a regular filter. We went sightseeing every day. We went to programs where they did customary dancing, sword fighting, kickboxing and bullfighting without the killing of the bull. We went for a swim in the Gulf of Thailand which is part of the South China Sea. The water was heavy with salt. The waves were pretty high. I would say about 5 feet. We would swim out to catch a wave and body surf back. It was easy but nobody told us to stop short of the beach. We rode the wave until we got into the shallows and pretty close to the beach. Once you hit the shallows, the sand and gravel raked across your chest, stomach, legs and feet. We didn't learn the first time and went at it again. This time the scratches got deeper and the salt water was burning. We went straight back to our hotel, changed and jumped into the pool. Bangkok was a lot of fun. There was a lot to see and do. Don and I bought civilian clothes for our return home and the food was excellent.

Temple in Bangkok

65 pound Boa in Bangkok

The three of us left Bangkok on November 22 and landed at Ton Son Nhut airbase that afternoon, the day Don and I had to report back to our duty station. All we had to do is hitch a ride on an aircraft going in our direction and hopefully get to Phu Loi and not miss the mail chopper going back to Lai Khe. We could not get a helicopter or fixed wing aircraft out of Ton Son Nhut. I told them I had another way out. We ran into a South Vietnamese Colonel near the MACV (Military Assistance Command Vietnam) compound driving a jeep. We asked him if he could give us a ride to Phu Loi. He said he could but we have to buy him a bottle of Cognac. We agreed and he took us to the PX (Post Exchange) to buy the Cognac. He would not take the bag because he could get in trouble. We climbed into the jeep and off we rode to the MACV gate. There we were able to drive straight out without guards and on to the two lane asphalt road. It took over an hour to get to Phu Loi. By this time it was around 1830 hours. He stopped at the front gate and he told us to "didi", which means to get going. There happened to be 2 MPs (Military Police) at this gate. The Colonel in the jeep drove off and right away they got tough with us. We were in civilian clothes and carrying canes we bought in Thailand. My cane at the head had a peace sign and Donalds had the middle finger sticking up. They took our canes and said it was against regulations and asked for our I.D.s. After about 20 minutes arguing with these MPs they let us walk in. Donald and I headed to the flight line, hoping to catch the mail chopper to get back to Lai Khe. No chopper was in sight so we went to the COs office and asked if we missed the mail chopper to Lai Khe. The Sergeant said it hadn't come yet and dusk is setting in. Then as last light a chopper was landing. We waited by the mail sack to see if it was our guys from Lai Khe. An officer and an crew member came in. We asked the officer where they were going. He said Lai Khe and we then asked for a ride back. He said yes but told us to grab some of the boxes. Don and I said to each other how lucky can we get. We made it back to Lai khe on time, helped with the mail packages, then we were told TOP (First Sergeant) wanted to see us. Right away he said we were AWOL (Absent Without Leave). We told him we just got back from R & R on our assigned day. We gave him our paperwork and he said you're lucky you're flying tomorrow. We went about our business to settle in for the evening as a Staff Sergeant approached us and asked us where have we been. We told him we just got back from R & R and showed him our paperwork. He asked us how come nobody knew about this. We said we have no idea.

CHAPTER 18

Back To Work (On My Birthday)

The next morning, after returning from R & R, was my birthday and only the gunners knew it. As I was suiting up to get on the helicopter someone in my platoon told the pilot and he went off on me. He said, "WHY DIDN'T YOU TELL ME IT WAS YOUR BIRTHDAY?!" I told him I did not think it mattered. He said, "no one flies on their birthday, it is bad luck!" I left all my gear on the chopper so who ever took my place was ready to go.

Me by one of our patriotic doors showing our loyalty

I went back to flying missions on November 24th. The pilot apologized for being so brash with me and explained the superstition. He said we have intelligence that a POW (Prisoner of War) compound was possibly located in the jungle near the Cambodian border. There was one mission before mine. They could not find the compound. The jungle was extremely dense. As we flew over the tops of trees, I caught a glimpse of a possible hooch and told the pilot. He told me to guide him close but not over the sight. There was a rifle platoon staged and ready to be inserted if we could confirm the presence of POWs. Our pilot called to C & C and the Cobra pilot to look to the west of our aircraft. The Cobra and C & C crews had binoculars to look at the sight from the altitude they were flying. We did not want to spook the enemy force guarding the American Prisoners of War. We did not know what they were capable of. We got a call back from the Cobra pilot saying he sees no movement in the compound. They guided us in for a closer look and discovered it was a POW compound but it had been abandoned. There was an overgrowth of vegetation on this site. Evidently the POWs had been moved some time ago. This was one of my biggest letdowns. I thought for sure we could have rescued our men. To make this compound unusable I threw white phosphorus grenades to burn the hooches and towers. The Cobra then rolled in on the burning hooches and toppled the guard towers with a minigun, rockets and an automatic grenade launcher blazing.

CHAPTER 19

Going Home

My time is short in country. I got my orders to go home on December 15. All the officers knew I had a short time left so they started giving me less and less missions. They figured as much as I flew combat missions my chances of survival became risky and being with a new guy or a short timer made it even riskier. They kept me flying until four days before going home. It would have been too boring to sit and do nothing. Three of the most experienced gunners went home within a week of each other. I was the last of the three to leave. My time came to leave A trp 3/17. I experienced many feelings and emotions. I really did not want to leave knowing we had inexperienced observers and gunners. I felt a responsibility for them. I questioned myself, did I give them enough training and information to survive in combat. Not only were their lives at risk, the lives of the other two crew members depended on him not to hesitate and "pull the trigger." Failing at either one could cost all three their lives. I was told the evening of December 12th to get my belongings together and be ready to go at a moments notice. The morning of December 13th I was to report to the flight line for my last ride out, taking me to Ben Hoa Airbase. At Ben Hoa is where you are processed out. You are put into a holding company until the plane you are assigned arrives.

Last ride out of Lai Khe

My flight was delayed another day. On the 16th They gave us the opportunity to call home. What a pain in the ass that was. You had to go through three different phone operators and every time you got done with your statement you had to say "over". You had to instruct whoever you called to say "over" when they were done speaking. You did this because there was a delay between conversations. I got my Mom at work and told her our plane was leaving on the 17th and I should be home on the 19th. You lose a day crossing the International Date Line . She said let me know when you get to O'Hara airport because she wanted to send a limousine to pick me up. I said no Mom I have to make it home on my own.

Headed for the freedom flight out ofVietnam

Before boarding the plane we had to go through customs. Once cleared our group was to assemble in a large conference room at Ton Son Nhut airbase that was connected to Ben Hoa. When our plane was ready we were told to board as a group so they could do a head count. We took off and when the Freedom Bird got over international waters our pilot announced that we are out ofVietnam and combat zone free which was met with exhilaration and relief. Our first stop was Hawaii to refuel and change crews. We had about a 2 hour layover. We then took off for Oakland, California. When we landed in Oakland, we were told to go to the debriefing room. It was a large hall within the airport. Once everybody was in, they closed and locked the doors. A Captain spoke first and welcomed us back to the states. Then in less than 5 minutes, turned the podium over to a Sergeant. The Captain left the room and our debriefing began. The Sergeant said, words I will never forget, "you are under federal protection for 90 days. If Jody has gotten into your life you can take care of this. (Jody is the guy who goes out with your girlfriend/wife while you are deployed) Just remember your time limit."That actually bothered me. That is why it is so vivid in my mind. We were released immediately after the Sergeants speech.

We were at Oakland International Airport, so I headed for the T.V. screen looking for the first flight out. I booked a flight with American Airlines to Chicago. There were five of us that flew in at the same time and were getting on this flight to Chicago. Once we boarded the 747 Jumbo Jet we noticed it was pretty empty. The 747 in 1971 carried 660 + passengers. On this 747 there were approximately 35 passengers. Once we were airborne the flight attendants asked everybody to move up to 1st class. The flight from Oakland to Chicago took about 4 1/2 hours. The five of us from Vietnam sat together and laughed our asses off while having drink after drink. The plane ran into turbulence and all of us had one drink in each hand. As the plane dropped our drinks would go up with our hands as if we were given a toast. Not one of us spilled our drinks and even the attendants were laughing with us. When we landed in Chicago the attendants had bottles of champagne wrapped in towels and handing them to us as we got off the plane. They thanked us for making the flight pleasurable for everybody.

When we got off the plane we went to the baggage claim for our duffle bags. I also had the opportunity to call my Mom. She asked me again if I wanted to take a limousine or cab home. Again I told her I have to make it home myself. As the five of us started to walk through the airport a Chicago Police officer said he would escort us out, we just looked at him and keep walking. There was a group of hippie looking people standing near the exit looking at us as we approached. Then they started walking towards us and I gave them a stern warning, "do not take another step." Then they froze in their footsteps. Then they made their move again and I said in a louder voice and more stern, "this is your last warning." They stopped and we walked by undeterred. When we got outside the Chicago Police officer said, "I have never seen anyone stop them as fast as you did." We thanked him and went in our separate ways. When Vietnam Veterans came home, people such as these, spit on us, called us baby killers, threw stuff at us and called us every other name you can think of. All I wanted to do is get home. I was not prepared for the changes I was about to face.

CHAPTER 20

Almost Home

Outside the airport I got into a cab. I told him to take me to the I.C. Station (Illinois Central), now called Metra. It is the electric train that ran from Randolph Street east of Michigan Avenue and the end of the line going south was Park Forest. Later on in the late 70s the end of the line was University Park. I had a pleasant talk with the cab driver. During our conversation he mentioned he had been in Vietnam with an artillery unit. He asked me what I did over there and said I see you have a C.I.B. (Combat Infantryman Badge). I told him I started out as a M-60 gunner or radio operator with a rifle platoon and later in my tour I volunteered to be a Scout on an OH-6 that worked with a Cobra and we were called Hunter/Killer teams. He told me he was a grunt when he first got to Vietnam and got wounded. When he was ready to go back to his unit he was transferred to an artillery battery. After a 50 minute drive, we pulled up to the front of the Prudential Building. I leaned over to look at the meter as he was resetting it. I asked him how much, he said this ride is on me. I was in shock, I said "well, here is a tip." He said I can not do that to a brother. I thanked him and headed to the train station. As I walked up to the ticket counter, the clerk asked where I was going. I said, "Riverdale." He handed me a ticket and started messing with papers on the counter he was sitting at. Usually you pay first. I asked him, "How much for the ticket?" He said, "It had been taken care of" and pointed at a man standing by the wall. He was dressed in an I.C. Railroad uniform. I went over to him and thanked him. He said, "Anything for a man in uniform." He told me he got out of the military in 1962 before Vietnam heated up.

I got on the train and the conductor never asked to see my ticket. I did not go into the seating area. I stayed by the door holding the hand rail. The conductor came to where I was standing and asked me if I was going home. I said, "yes, I am." Then we talked a little more and he told me his brother went to Vietnam. The conductor left to do this rounds. As I looked out the window at the scenery that passed by, I remember how I loved to ride this train to downtown Chicago. The conductor was passing by me going through the cars and said, "Riverdale is two stops away, if I don't see you again - Good Luck!" I got off at my hometown train station and watched the train pull away. I looked at Riverdale Park to the west and to the houses and apartments to the east. Nothing had changed in Riverdale from what I remembered. It is as if time had stopped. 138th Street was quiet - 2 or 3 cars driving by the south exit of the train station. I am now a mile from home. I started to walk west with my duffle bag slung over my right shoulder. A car pulled up and the guy driving rolled down the passenger window and said, "Hey soldier - do you need a ride?" I said, "No." He asked me how far I was going. I told him I have about a mile to go. He said, "Let me give you a ride." I told him I have to do this on my own. He finally said OK, rolled the window up and drove away. Then in about a hundred feet he stopped along the curb. He had the window down and said, "Come on, let me give you a ride." Now I have walked about a block and I was getting cold. The sun is starting to go down and the temperature is beginning to fall. I left Vietnam where the temperature was 90 degrees and rising. I am near my home and it is 29 degrees and falling. I figured what the hell, take the ride. I got in and said I do appreciate this. He told me his brother would probably kill him if he found out he did not give a soldier a ride. He took me home where my Mother, Grandparents and Uncle were waiting. I hugged my Mom, Grandmother, Grandfather and Uncle. A very warm welcome home from a loving family. Before I came home I asked for a spaghetti dinner. It was on the table with Italian sausage. I changed my clothes and we all sat down to eat. We had a spaghetti dinner every Sunday at 12:30pm. It was the best meal I had in a year.

1st hour home

CHAPTER 21

Adjustment

My first night home was an adjustment in itself. Every noise and smell put me on high alert. My dog did not know me. I figured it was just as well, because I was going to have to leave again in 39 days. Her name was Blackie, she was an excellent hunting dog. When I left for basic and A.I.T. (Advanced Individual Training) she was 18 years old. I did not have to train her to hunt. She was a natural. She would never go more than 20 - 30 feet in front, running back and forth in a zig-zag pattern, looking back occasionally, keeping her distance. She was amazing with pheasants. She would point with one front leg up freezing the pheasant. On command she would flush the bird for me to shoot.

On December 20th I called two guys I met in Vietnam that were from my area - a town east of Riverdale called Dolton. We all knew the same people but never ran into each other. I was hoping they made it home. I called Randy first and he answered the phone. Randy and I were in the same unit, A trp 1/9 at Tay Ninh before they stood down. We agreed to get together. Randy said tonight is really the only night he has free. I said, "How about if I give Roy a call." Randy said, "Hell yeah - see if he can come out tonight." I called Roy and he, as well as I, was happy we made it home. Roy agreed on coming out tonight. Randy offered to pick both of us up in his new 1971 Ford Torino Cobra. Randy picked us up around 3:00pm. Our First stop was a bar in Riverdale called Ivanhoe Liquors. We had some beers and hamburgers. We were joking and laughing with the bartender. We were discussing going to another bar in Riverdale called the Fin & Feather. A guy sitting at the bar overheard us say we were going to the Fin &

Feather and asked for a ride. We looked at each other and said sure you can come along.

We left Ivanhoe Liquors and when we pulled into the lot at the Fin & Feather, two Riverdale Police cars pulled up blocking us in. Frank the police officer said with his gun pulled, "get out of the car with your hands high." Then he told the four of us to turn around and put your hands on top of the car. Frank told the other officer to frisk us. As he was frisking Roy, he found a small pocket knife which he handed to Frank who put it in his pocket. Roy started to turn and said, "we just got home from Vietnam." Just as Roy finished his statement, Frank hit him with the butt of his pistol on his right shoulder. The knife Roy was carrying was given to him by his Grandfather which he carried with him throughout his tour in Vietnam. Franks partner searched our car and told Frank there is nothing there. Frank then said, "Book them!" He asked Frank on what charges? Frank said, "I don't know - just book them." Franks partner grabbed him by the arm and pulled him to the side. During their conference you could hear them arguing. Frank came back over to us and said, "I'm going to let you go but get out of this town and don't come back." I said sarcastically, "I live in this town." He said, "Do you want to get smart with me?" I told him, "not right now" and we just stared at each other. He knew what I meant. I will run into him somewhere in town when he is out of uniform. His partner grabbed him by the arm and said, "Lets go." We went into the bar and Casey the owner was working. He saw what was going on. We told him our side of the story. He was another one that did not like Frank who had a reputation of being an asshole. We told him he took Roys Grandfathers knife. A lot of lawyers, politicians and cops frequent the Fin & Feather. Casey asked his Cook Country cop friend to talk to us. He told us to file a complaint with Cook County. He said they will arrest him. Franks reputation was well known. The county cop told us to go to the police station and tell the chief what your intentions are.

We took his advice and headed over to the Riverdale Police Department. We asked to see the chief and the officer that was with Frank said, "You're the guys at the Fin & Feather." I said, "We are and we want an apology, Roys knife back and to file a complaint." The officer said, "Hold on the chief needs to hear this." They came walking out and said, "I hear you want a knife back." I said, "yes." He said to the officer standing next to him, "you were on the stop - give them the knife." The officer started to reach in his pockets and I said right away, "He handed it to Frank and Frank put it in his pocket." The chief said, "we are going to have to call him in." I told the chief we also want an apology from Frank. The chief said, "you're asking for a lot." I told the chief, "we did nothing to be stopped, we were trying to have a good time when these two officers told us to get out of our car at gunpoint. We just got home from Vietnam and Frank hit Roy with the butt of his pistol in the right shoulder. If Frank refuses to give Roys Grandfathers knife back and apologize to us we will sign a complaint with the Cook County Sheriffs Police." The captain said you might get your knife back but an apology I'm not too sure of. I did all the talking and said, "then it is up to him on which way we go." Frank did not make us wait long. He said, "whose knife is it?" Roy said,"It is mine." Frank reached into his pocket and handed the knife back to Roy. Frank said, "Where is the fourth Guy?" We told him we met him at Ivanhoe Liquors and he overheard us say we were going to the Fin & Feather and asked us for a ride. When we walked into the Fin & Feather he went to the other end of the bar to talk to people he knew. Frank then apologized to all of us and said, "The bartender at Ivanhoe Liquors called and said he had a gun." She overheard you say you were going to the Fin & Feather, so that's where we intercepted you. The room got quiet, the captain then said, "Are you guys satisfied?" We looked at each other and said yes and then the chief said we can all go about our business. This stress and anxiety took a lot out of us. We just came home from hell and we did not need this.

CHAPTER 22

Christmas

December was hunting season and my uncles Roy and Isadore asked if I wanted to go hunting. Without a thought I said I would. We went out December 23rd and it was cold. I took my 12 gauge shot gun and #8 shells. When we got to the farmers land that we hunt, we took the normal path as always, getting to the corn field. This was strange to me. I'm carrying a weapon and scanning the area for movement rather then looking for rabbits and pheasants. I had to convince myself we are hunting wild game. We hunted in the cold for 2 hours and decided to pack it up. Not one bird or rabbit was seen. On our way home I'm thinking to myself you can not kill anything any more.

Christmas Eve and Christmas Day came. I was very anxious knowing my aunts, uncles and cousins were coming over Christmas Day as they have every year. I was very uncomfortable with all of these people packing into the house. I had to pretend I was OK and keep smiling. Everybody was very happy to see me. I thanked my aunts and uncles for the goodie packages they sent to me in Vietnam. I'm hearing noises outside that the rest must take for granted. Every once in a while I would look out the window or open the door and look through the storm door until it fogged up. After eating I told my Mom I need to get out of here for a little bit. My cousin Joe and I, who is also the Vietnam Veteran that I met at the Rockpile in Vietnam, went for a ride. He said to me when he got in the car I needed to get out of there too. We went to our friend Tommys garage. This is the place where everybody hangs out. After a couple of hours it was time to go home. I dropped Joe off at his house in Calumet Park, Illinois. I went home and by that time everybody was gone.

The next morning the phone rang. My Mom answered the phone and said the call is for you. It was Candy, I felt no animosity towards her and we talked for an hour. Then she asked if she could come over. Excitedly I said yes. In about a half hour she drove up in a real nice Mustang. My Mom told me she had a job so when I went out to greet her I asked if it was her car. She said no it was her Moms and her parents had gotten a divorce. She asked if I wanted to get something to eat. I said yes where would you like to go? She said how about Dons - our favorite place. Dons was a fast food restaurant that had the best Italian sausage with a red sauce and hot sport peppers. That was a treat for me, two times in one week, Spaghetti dinner and Dons Italian Sausage. After we got done eating, Candy said my Mom would like to see you. She drove us to her house and I said hi to her Mom. After talking for about an hour, it was getting late and Candy had to work the next day. She drove me home and asked if she could come over tomorrow. I said yes. We saw each other every day and I never asked about the guy she was seeing. She drove me around to see my friends from town. They were still in the same place as when I left. Everybody seemed changed. Nothing was the same anymore. I grew up in one year and they were still doing kid things and nothing seemed serious to them. When Candy picked me up on Saturday and said she did not want to go into my Moms house. She wanted to go out and get something to eat and said she had tickets to a play. I said that's fine with me. We had a lunch/dinner prior to the play. When the play was over she said she wanted to go home to change. We pulled up to the parking space behind her Moms townhouse and she said, "I want to let you know I am engaged and my fiance is on vacation with his parents and coming home tomorrow." These words I will never forget. She also told me the car is hers and her fiances. I started to cry immediately. I got out of the car and started to walk. She said, "where are you going?" I said, "home." She said, "Let me give you a ride." I said, "no" and walked away. At this point I never cared to see her again.

On December 31, 1971 I met my future girlfriend/wife. My buddy Emmett and his girlfriend Diana pulled up by my Moms house and we talked for awhile inside his car. Then Ray pulled up behind us, a friend of Emmett and I, Ray was with this girlfriend who happened to be Dianas sister Sandy. Emmett left so I got in the car with Ray and Sandy and talked with them for while. I asked if they were doing anything for New Years Eve. They said no and shortly drove away. Sandy and I barely broke eye contact the entire time. She was a very beautiful woman.

On January 2, 1972 Emmett called and asked if I wanted to go out with him and Diane. I had nothing to do so I said come on and get me. We went to a bar in Calumet City called Gumps. It was good to get out. We went out the next couple of nights to the same place. Then Diana asked if I would like to go out with her sister Sandy. Diana called her sister by her middle name Jolene. I told Diana I'm only going to be home another ten days. Diana said that's ok I will call her and see if she would like to met you tonight. Diana called her sister from the bar and asked if she would like to meet me. Sandy said no way do I want to meet any more of Emmetts stupid friends. Diana did not tell me that she said that. Diana said when Emmett takes me home, you can meet Jolene (Sandy). When we pulled up to Dianas house she said come on in. Here is Sandy sitting on the living room floor in her robe putting curlers in her hair. You could see she was pissed, said hi and left the room. The next day Diana said her sister would go out with me. Somewhere around the fifth or sixth of January we started to date. I had no idea where to take her so she recommended The Old Towne Tap in Crete, Illinois. There Sandy ordered whiskey sours and I struck to beer. We got a pizza which was very good. Many people knew her at this Bar Restaurant. Then all of a sudden a certain song came on and Sandy said I will be right back. The crowd seemed to have waited for her to get to the head of this dance where they go up and down the aisles and between the tables similar to the Bunny Hop. Everybody was holding on to the hip on the person in front of them. There had to be 30 to 40 people in this dance spectacle. She was very outgoing and I am still trying to figure out how I fit into society.

It was very difficult being around all these strange people. We went there the next night also. They had a raffle and she won a bottle of Cold Duck Sparkling Wine. We did not stay long. We went to my house and parked near the garage and opened the Cold Duck. We drank about half the bottle, put the plastic cork back in, set it on the back seat and began to hug and kiss. It was a cold January night and the heat was blaring in the car. It did not take long and the cork popped out with a loud bang. Immediately my survival skills kicked in. I shoved Sandy down to the floor of my car. I covered her and put the car in gear. I was able to peak over the center of the dashboard as I drove off. Sandy is yelling, "WHAT IS WRONG WITH YOU?" I said, "someone is shooting at us." She said, "It was the cork on the Cold Duck." Sandy then said, "It is time for me to go home anyway." I apologized to her all the way home and explained I just got back from a combat zone and swore someone was shooting at us. When I dropped her off I asked if she would see me again. All she said was I will call you tomorrow. I waited the next day and she finally called after getting off work and asked if she could pick me up. From that day forward we were never apart if at all possible.

On January 13, 1972 Sandy took me to the airport and waited until I boarded the plane to Colorado Springs, Colorado. I had to report to Ft. Carson on the 13th. We hugged and kissed and she cried as I left to board the plane. I did not expect her to wait for me nor did ask her to. All Sandy wanted to know is how long I was going to be gone. I honestly did not have an answer for her. From my understanding they were giving guys a 6 month early out. That was all I could hope for. I told her I would try to call when I could.

CHAPTER 23

Fort Carson

On my arrival at Colorado Springs I was not 100 feet into the terminal when a Colonel approached me and said, "You need a haircut solder." I looked at him and kept walking without saying a word. He yells, "I am talking to you soldier." I did not give a shit. A Sergeant caught up to me and talked to me like a human being. The Sergeant said he is only doing his job and understands I just got back from Vietnam and he could see by my medals I was a combat veteran. He then asked me to come back with him to the Colonel. I said ok let's go. The Colonel asked me for my unit. I told him I do not know what unit I'm going to be with because I am reporting for the first time. In his little pocket notebook he wrote my name, rank and serial number and said, "I will see you at the base and if your hair is not cut I will give you an article 15." I looked at him and said, "Is that all sir?" He said, "yes." I turned and left and found out my bus was not coming for another hour. I figured I better get a haircut. The Colonel and Sergeant were gone so I went into the barber shop. The barber knew exactly what kind of haircut I needed.

Once I got out of the airport I could see how beautiful Colorado was with majestic mountains with snow covered peaks. I got on the bus and it took me to the bases main gate. I checked with the MPs at the gate and showed them my paperwork. None of them knew where I was to report. I started to walk onto the base and in about a half mile I saw some guys in their dress uniform as I was. I asked if I was in the right spot, showing them my paperwork. They said they were going home and maybe check in at the office down the road. I walked in there and showed them my paperwork and they then said once I check in to come back here. I said,

"OK where do I check in?!" They sent me back to where the guys were going home. I finally got to check in and by this time, I was late. Luckily my paperwork was wrong. I told them I have been trying to figure out where I was supposed to be. No problem at this point, then the clerk calls the Officer on duty over so he could sign off on my haircut. The Colonel at the airport had send them my name to make sure I got a haircut. I was not used to this chicken shit stuff in Vietnam. The very next day I was told to polish my boots. That meant going to the PX and getting polish. Now I was getting familiar with military life stateside.

I had to report for roll call every morning at 0630. When roll call was complete we were released to go to breakfast. We had to report back at 0800. Five guys at a time were called into the office and the rest of us were either released for the day or we had to do a police call (pick trash off the ground). After about four days almost everybody was sent to their duty station for the remainder of their commitment or career. I was finally called. The Captain said that I had an impressive military record. The Captain turned me over to a Sergeant that offered me 15 thousand dollars to re-up. I told him I would consider it if I could have the same job I had in Vietnam. He then went into detail about being a Scout. He told me straight up they have no more use for AeroScouts, but he would check and get back to me. He let the captain know what we discussed and I was released for the day.

I would walk across the street with some of the guys from my holding company. There you could get pizza or a hamburger and from a vending machine a 20 ounce cup of beer for 25 cents. The next day we did the same thing going across the street. I finished my beer then my hamburger was ready. I got up to get another beer, sat down and started to eat. The Sergeant walked in and said, "Gregory, the Captain wants to see you." I told the Sergeant I would be right there. He said, "he wants to see you now." I said, "OK, as soon as I am done." He said, "I can not go back without you, can you hurry it up?" I finished my hamburger and gave my beer to the guys I was with. The Sergeant walked with me to the Captains office and he explained to me that the job I had in Vietnam as a Scout was longer used. They did have operations going on in South America but he could not guarantee that is where I would go. I told both of them with no guarantee I am not interested in re-upping right now. The Captain then said, "we have openings in Washington D.C. as an honor guard." I told him, "no thank you Sir." He said, "you need to do something for the remainder of your time. You will be working with horses along with protecting the President." I asked about the 6 month early out. The Captain and the Sergeant looked at each other in disbelief. The Captain asked me to reconsider because once he starts the paperwork for my early-out all offers are off the board. I said, "I understand that Sir, I just want to go home." He said, "OK, the Sergeant will get you when it is time to sign the papers."

For the next few days, after 0630 roll call, five of us getting an early-out headed to Colorado Springs every day. We had to be back by midnight. We found a hardware store that sold rifles and rented them to fire in the ally. Being that we were in the military all we had to do is buy a box of ammo. The side door of the hardware store opened to the alley. They had bales of hay stacked at the end of the dead end alley. We were told to shoot adjacent to the door and not behind in case someone walked out. Shooting at targets was not the same to me without having to worry about being shot back at. I only fired a couple of rounds and gave the rest of my ammo to the two guys I was with. After about twenty minutes, we left the hardware store and looked for a place to eat. Right outside next to the alley was a bar and in the window it said "Worlds Best Hamburgers." We walked up to a crowded bar and ordered beer and a hamburger each. The only place we could sit was against the wall with a shelf where we could set our beers and eat. In a little over twenty minutes our hamburgers were ready. As we ate we all agreed it was the best hamburger we ever had. We made it our place to eat every time we went to Colorado Springs.

The next couple of days we ventured out. We went to Broadmoore Ski Resort because we were told women skied in their bikinis. The ski slope looked very nice and there were very few skiing but no women in bikinis. We could not go too far because of our midnight curfew. The last bus left Colorado Springs at 2200 hours.

The last three days I spent on base I called Sandy and I told her I would be home soon. On the 25th I was called into the COs office. The Captain and Sergeant were sitting behind the Captains desk. I knocked and the Captain said, "Come in and have a seat." This informality was odd to me. I sat down and the Captain started by asking me if I gave staying in the military any more thought. I told him I have a job waiting for me back home. He did everything he could to try and persuade me to stay in the Army. All I wanted to do is go home. He said my paperwork will be ready in the afternoon. The Sergeant and Captain had a bet going that I would stay, so my separation papers were not ready. They said they would call me when they were ready. There were only three of us left in the holding company. At about 1500 hours I was called into the COs office. A clerk was now taking my information. The Captain walked in and said if I ever change my mind to call him because the Army could really use a soldier like me. I told him I would because who really knows what the future had in store for me so I did not burn any bridges.

CHAPTER 24

Seperated From Active Duty

On January 26, 1972, I called my Mom from the airport to let her know I was out of the Army and coming home. At the airport I bought a stand-by ticket. As I walked through the airport I saw Delta Airlines had an earlier flight leaving 20 minutes sooner. I walked up to the desk and asked if they honored other airline tickets. She said yes we do. I only had 10 minutes to catch this flight. Once boarding and seated in the plane I gave a sigh of relief that I'm finally on my way home. I will never forget this flight. The seats were plush and only 2 seats on either side of the aisle. It had a foot rest you could pull down from the seat in front of you. I had a steak dinner and a cocktail on my final flight home.

As soon as I got home I started to see Sandy right away. She worked for Illinois Bell Telephone in Harvey, Illinois Monday through Friday from 8am to 5pm . We enjoyed many of the same foods and ate out almost every night. When I was home on leave I introduced her to Spaghetti dinner on Sunday at my Grandparents house where I lived and grew up. Everybody loved the Spaghetti dinner at 12:30 sharp every Sunday. Many of my friends or relatives would show up.

The excavator I worked for before the military was about to roll into full operation in February. I stopped in and saw Ray the owner of Lotz Excavating. Ray was a very patriotic gentleman and appreciated our men in uniform. We talked for a good 45 minutes when he asked me if I was ready to come back to work. I asked if I could have an extra week before coming back. He said if you want 2 or 3 weeks to let him know. I said I will keep you informed. I needed this time to buy a car. I went to the car dealership where my Mom had bought her last 3 cars from. The company was Fiedler Motors in Blue Island, a family owned business where I bought my first car in 1969, a Dodge Polara with a 413 Hemi Wedge. Going back this time I found a real nice car, a 1971 Green Metallic Dodge Charger with a 383 engine.

I waited until the end of February to go back to work. Before my draft into the army I was a 905 union mechanic apprentice. I worked nights getting trucks, heavy equipment and smaller power hand tools ready for the next days job. It was a good job, I worked alone. One day the boss, Ray, came up to me and asked if I wanted to drive a truck. I told him I would, this way I would be working days.

I came in early every day to practice driving and backing up the lowboy. The lowboy is a semi tractor/trailer combination. I got my CDL within one week of when my boss asked if I wanted to drive a truck. At this time I joined Teamster Local 731. The timing was perfect. A guy was fired and I was put on the 4 wheel dump truck with a tilt bed trailer. On this job I also labored for the operator that ran the machine I hauled. Many times I would have a dump truck dump gravel where I needed it and I would tell the operator to finish what he was doing. I would then level the stone by hand.

Every job I had with Lotz Excavating had limited contact with other employees. After a year I moved on to driving the the lowboy making 7 to 9 moves a day. Usually I would get my orders in the evening and start my day anywhere from 3am to 5am. My days were 12 hours 6 days a week. Some of the guys were jealous of my pay. The operating engineers complained the most. They did not like that I made more money than them. I worked 12 hours to their 8 hours. I cornered each one that complained. I told them they could have had my job and to pay attention to what they are doing and keeping their nose out of my business. From my military experience I did not appreciate the back stabbing and talking behind my back. In the military if you have a problem with someone, you tell them to their face, none of this shit thinking it will not get back to you.

Sandra and I saved enough money to get married. On May 19, 1973 we were married in the Assembly of God Church by Sandra's Grandfather in Harvey, Illinois. We paid for our own wedding because we knew our parents did not have a lot of money to spare. Our wedding reception was at the VFW in Hazel Crest, Illinois. We had 186 people attend on the list and some I told to come because I forget to invite them.

Me, my wife Sandra, our son Daniel and our daughter Jill 1985

Lotz was a great place to work for. I worked alone and had limited contact with the boss. It was high energy and somewhat dangerous being around heavy equipment, loaded trucks and deep digs at times.

Sandra and I bought a house on April 1, 1975. Sandra and I had 2 children, Daniel born May 1975 and Jill born May 1978. I was hard on my children. Discipline was foremost for me because of my military background. My combat experience made me think that if you did not listen to me you were going to get hurt or die. I felt a big burden to protect Sandra and my children. I practiced throwing an axe and knife. I trained in Vietnam throwing a knife into a rubber tree right outside my hooch.

I became very proficient. I also had my shot guns from when I used to hunt prior to military service. I bought a 9 millimeter semi-auto pistol and a 357 revolver but never shot them. I was already capable of deadly accuracy with the pistols.

I met a guy named Bob at Lotz Excavating that was in the Army Reserves. We hit it off pretty good. We would scrap for iron, aluminum, brass and copper. We would repair semi-truck tires and large tractor tires. Paint heavy equipment, truck tractors and semi-trailers. We would also re-deck semi-trailers. All of this being done after hours. We made a lot of money on the side. We were working hard for Lotz, me being a lowboy driver and Bob an operating engineer and for ourselves. We talked and decided we need to go into business together. We checked out different businesses, most of which I could not afford. Then Bobs Dad gave us a suggestion and we followed up on it. We made a phone call to the company in San Mateo, California. They told us the franchise is becoming available in our area and if we like we can go to Indianapolis, Indiana to see how the process is done. This company was informed that we were coming down. We went there on a Friday and observed what the process involved. They were structurally repairing a poured concrete parking garage by epoxy ejection. After a three hour drive back home on Sunday we decided to make the move. I talked to my wife, Sandra and she was very supportive. Bob and I went to our boss at different times to tell him we are quitting. I gave Lotz a three month notice that I was going to leave. Bob gave them a one month notice because he was not working anyway due to the weather. Bob got all the necessary paperwork in to incorporate and the different tax IDs so that when I came in we were ready to go to work.

As time came near for me to leave my job I kept asking to train someone for my lowboy. Every week for a month I would ask to train someone. May 11, 1984 was officially my last day with Lotz. I went into the office and thanked them for taking care of me and it was a pleasure working all the years I was there. They looked at me and said, "you are really leaving?" I told them I have been asking to train someone. Evidently they did not take me serious. It was the top paying job in the company. Monday they put a driver on my truck and on the way back to the yard taking a turn at the corner an endloader slid off the lowboy deck and on to its side. For the 14 union credit years I drove this truck I only lost 2 machines. Neither one landed on its side, both landed on one track so it was easy to put back on and tie down.

May 15, 1984 was the official start date of our new business in Illinois. Bob and I started out getting some business in the residential area of repairing basement foundations to waterproof cracks. The first year in business was the toughest for my family. I went from making over $40,000 in 1983 to 7,500 in 1984. Thank god for my mother helping us out.

My mother Helen

The second year became more profitable. In our second year our first commercial account was with the U.S. Army Corps of Engineers, repairing the Lock and Damn at the Lock Port Locks in Joliet, Illinois on the DesPlains River. The same year we started working for the John G Shedd Aquarium. Then in 1988 we started working for the Chicago Park District at the Grant Park North underground parking garage. We were about 4 days from completion at the Grant Park Garage and in our last area on a ramp. I'm working off of a ladder when suddenly I heard what sounded like AK-47 gunfire. Someone was using an air hammer to break concrete somewhere in the parking garage. I broke out into a sweat and shook. This happened everyday until we left. I would make an excuse to use the bathroom to get away from the sound. I could not tell my partner what was going on. I knew I could make it another day until we were out of there.

The park district job ended and I did not feel completely relieved. It stayed with me for days. Shortly after we went to the Metropolitan Sanitary District of Greater Chicago in Stickney, Illinois. There I was working on cracks between pipes when all of a sudden I heard a "whop". It sounded like a mortar round going off. I shook and went outside. I Came back in and there was a "whop" again. So I walked further into the tunnel and saw 2 electricians. I asked them if they heard any "booms". They looked at me funny and said no. It was happening 3 days straight at about the same time every day. I am extremely on edge by this time bringing back memories of sounds from Vietnam. The next day I caught the electricians letting the lid of their gang box slam shut and with the echo in the tunnel it sounded like a mortar going off. I asked them nicely if they would set the lid down without banging it. It told them I was working between pipes and it was making me jump. I could not tell them the truth that I was having flashbacks. The very next day I'm still wired and they slammed the lid down on the gang box again. This time I was pissed. I walked down there while they were eating lunch and said, "I asked you nicely not to slam the lid on the gang box. If you do it again I'm going to shove the gang box and you outside until we finish." I was extremely on edge and neither one of them said a thing and left. That bothered me that I got so angry. The only thing I knew was that it started with what sounded like enemy AK-47 fire and then mortar rounds impacting. I left the job at 1530 and my ride home was about an hour away. Something was wrong with me and I was scared and knew someone could get hurt saying the wrong thing to me. I got off the expressway and called the VA Westside Hospital in Chicago, now called Jessie Brown. I talked to a nurse and told her I felt like a time bomb ready to go off. She asked me to hold on. I waited and a Psychiatrist got on the phone and I told him the same thing. He asked where I was at and I told him. He said you are not that far away. Get in here now. I went directly to the hospital and met with the Psychiatrist. From that day I started therapy with a Clinical Social Worker three times a week. I was diagnosed with chronic PTSD related to my combat experience. My therapy is continuing to this day with group sessions, individual and other evaluations for my combat related Post Traumatic Stress.

The time I spent on active duty in the United States Army changed my life forever. It starts on day one. The humiliation of getting naked in a line with a bunch of other guys and having to bend over and spread your but cheeks. Getting off the bus where your training begins and Drill Sergeants are yelling in your face. He then begins to tell you he is now your mother, father, brother, sister… at this point you know everything in your life is about to change. As our first instructions are being told to us I remember what my cousin John, a Combat Marine, who was given last rights in Vietnam from shrapnel wounds, told me, "Learn all you can in boot camp, it can keep you alive and do not volunteer for anything."

When you take your first step onto Vietnamese soil it all becomes real. Helicopters with armed troops, other helicopters with artillery, supplies and water bladders attached by cable heading to the front lines. The smell of jet fuel (JP4) from helicopters and jets taking off. The heat takes your breath away when you are getting off the plane. The smell of burning shit in cut-off 55 gallon drums from the latrines.

Many emotions affect you in different ways. Smells from burning wood in the jungle is a good indicator that the enemy is close. It peaks your hypervigilance. When in the jungle the heat and humidity are tremendous. You are not only looking for movement, booby traps, trip wires, but also snakes, spiders, scorpions and other critters that can kill you.

When you are loading on to a helicopter to take you to a LZ (loading zone) your apprehension goes up. On the way to the LZ you think about dying. You say to yourself, "Is today my day or will I luck out again?" You're scared the whole time. I was told to walk point for the first time. You are the first in front of the main unit and you are the first to make contact if you run into an unexpected enemy force. After being in a rifle platoon for three months I volunteered for Scouts. Our job was to fly low level hunting the enemy. We were always put into hazardous areas where there was known enemy concentrations or movement. When we came up on them it was fast, sudden and extremely violent. Depending on the pilot it could last seconds to a minute fighting for your life and your crews life. Your adrenaline kicks in to new heights. You just fought for your life then when it is over you can feel your body shrink and your sphincter release. This takes about 40 minutes after the battle is over. You do 3 to 5 missions a day lasting anywhere from an hour and a half to 2 hours on the deck at tree top level or the tops of brush. The worst is the silence when it is all over. It becomes surreal. You go from extreme violence to silence and motionless bodies lying on the ground. This is where the "thousand yard stare" comes from. Basically you are still alive and your brain is still trying to comprehend what just happened here.

So many emotions and feelings run through your brain and body, silence is the loudest noise you can hear. It is the worst. We are hiding emotions. Some of the guys get angry and start kicking things or punching trees or walls. All you have is your comrades. You laugh together, you cry together and make sure they are prepared to keep moving because the mission does not end. When one of your own gets shot you are either pulling security for rescue/recovery or help treat and evacuate. There is no time for grief.

All the sights, sounds and smells go home with you. For years I would drive past tree lines that looked way too familiar looking for movement. My chest would tighten up and my breath would become erratic. One day I was going fishing with some buddies at their favorite fishing hole, when all of a sudden I realized I am walking on a well traveled trail with a red colored clay. I started to panic but I could not let them know what was happening to me so I said, "I have to take a leak, I will meet you at the creek." I went off the trail about 20 feet. I just stood there breathing hard and getting my composure together and convincing myself I am safe here. Sudden noises, fireworks, someone touching you when you're not expecting it are triggers. I do not go to malls because of the amount of people that are there. In a war zone you never congregate because that is where the enemy would throw a hand grenade or lob a mortar. One time going to a check point we stopped at a village and the guys were opening their sea rations for the candy to give to the kids. There are only 15 of us and 10 of them are in a group handing out candy. If they got killed I am pretty much on my own.

Once you have endured combat over and over again, you are scared every single time until you resign yourself to dying. You don't relax you just say if I am to die please make it fast. Every day you survive is a blessing. War goes against all your moral values you were taught as you were growing up. You were always told not to kill.

"You have never lived until you have almost died. Life has a certain flavor the protected will never know"

Individual Awards

National Defense Service Medal
Vietnam Campaign Medal w/60 Device
Vietnam Service Medal With 3 Bronze Stars
Bronze Star Medal
C.I.B. (Combat Infantryman Badge)
Air Medal With "V" Device With Numeral 2
Republic ofVietnam Gallantry Cross With Silver Star
Army Aviation Aircraft Crewman Badge

Unit Awards and Citations Issued by the Republic ofVietman

Gallantry Cross with Palm Citation Badge
Civil Actions Honor Medal
First Class Unit Citation Badge

Special Thank You

A special thanks to my friend and Editor Gail Patricia Whitright. Miss Gail Whitright is a teacher, an American Patriot and founder of PROPOSITIONUSA.COM - The goal of which is to help Veterans and all Americans. Gail's heritage dates back to the American Revolutionary War where her Grandfather 6 Michael, Grandfather 5 Peter Johan and Uncle George Whitright served in the Pennsylvania Militia - the first United States Marines from 1775 to 1783. After the Revolution Peter Whitright was given land in Ohio where he joined the Ohio Militia and fought in the War of 1812.

Gail is a member of the Daughters of the American Revolution. She also has family members that fought in the World War 1, World War 2 and the American Civil War. Miss Whitright has dedicated much of her life helping Veterans. She is an escort for the Iwo Jima Memorial in New Britain, Connecticut and participated in the first "Welcome Home Vietnam Veterans" in Cheshire,Connecticut as well as Shelton Connecticut. Please help Gail in her endeavors to helping Veterans by joining her on PROPOSITIONUSA.COM which is on Facebook.

gaiwhi860@aol.com

Made in the USA
Monee, IL
10 August 2026

58664821R00069